A DESCRIPTIVE CATALOGUE OF THE SANSKRIT ASTRONOMICAL MANUSCRIPTS PRESERVED AT THE MAHARAJA MAN SINGH II MUSEUM IN JAIPUR, INDIA

Compiled by David Pingree

From the Notes Taken by Setsuro Ikeyama,
Christopher Minkowski, David Pingree, Kim Plofker,
Sreeramula Rajeswara Sarma, and Gary Tubb

American Philosophical Society
Philadelphia • 2003

Memoirs of the American Philosophical Society
Held at Philadelphia
for Promoting Useful Knowledge

ISBN: 0-87169-250-3
US ISSN: 0065-9738

Library of Congress Cataloging-in-Publication Data

Pingree, David.
A descriptive catalogue of the Sanskrit astronomical manuscripts preserved at the Maharaja Man Singh II Museum in Jaipur, India / compiled by David Pingree from the notes taken by Setsuro Ikeyama ... [et al.]
p. cm. -- (Memoirs of the American Philosophical Society held at Philadelphia for promoting useful knowledge, ISSN 0065-9738 ; v. 250)
Includes bibliographical references and index.
ISBN 0-87169-250-3 (cloth)
1. Hindu astronomy--History--18th century--Catalogs. 2. Jai Singh II, Maharaja of Jaipur, 1686-1743--Catalogs. 3. Mahārājā Mānashiṃha Pustaka Prakāâá--Catalogs. 4. Manuscripts, Sanskrit--India--Jodhpur--Catalogs. I. Pingree, David Edwin, 1933- II. Title. III. Memoirs of the American Philosophical Society ; v. 250.

Q11.P612 vol. 250
[QB18]
081 s--dc22
[016.520'954]

2003061286

Table of Contents

PREFACE

In the mid 1990s the Seminar on Science of the Dharam Hinduja Indic Research Center at Columbia University undertook as its research project an investigation of the astronomical activities initiated and carried out by Sawāī Jayasiṃha, the Mahārāja of Amber and later of Jayapura in Rajasthan between 1700 and 1743. We chose this project both because there is a vast amount of unpublished manuscript material, in Sanskrit, Arabic, Persian, and Latin, relevant to the Mahārāja's activities, some of which we already had access to through photocopies, but most of which we knew still to be preserved in Jayapura itself, and because the efforts of Jayasiṃha to assimilate to Indian usage first Muslim and then European astronomy and the resistance he encountered promised to serve as a model case for approaching the far more complex process by which European science gained its dominant position in India in the nineteenth century.

With the financial support of the Dharam Hinduja Indic Research Center, we were able to spend the month of January in 1998 in Jayapura cataloguing the Sanskrit astronomical manuscripts in the Maharaja Man Singh II Museum. Our team consisted of six scholars: S. Ikeyama of Brown University, C. Minkowski of Cornell University, D. Pingree of Brown University, K. Plofker of Brown University, S. R. Sharma of Aligarh Muslim University, and G. Tubb of Columbia University. Each of us spent between one and four weeks working in the library of the Museum, describing the manuscripts and recording the information needed to identify the texts accurately and to determine the origins and histories of the manuscripts. The descriptions were placed in the hands of the present writer, who has compiled from them the present catalogue. We also were able to photocopy those manuscripts which seemed most important for the completion of the "Jaipur Project" as well as others whose texts were particularly rare or unusual. The original photocopies are with Prof. Pingree at Brown University.

In the course of this work we were assisted by many people, whom we are pleased to thank here. First we must express our profound gratitude to the present Mahārāja of Jayapura, Sawāī Bhawani Singh, for graciously allowing us to examine the manuscripts. We are also very grateful to Dr. Yaduvendra Sahai, the director of the Museum, whose kindness and hospitality increased immensely the pleasantness and profitability of our stay in Jayapura. We are also most appreciative of the patience and helpfulness of the library staff.

The staff of the Dharam Hinduja Indic Research Center at Columbia University also deserves our abundant thanks, especially the former director, Mary McGee, and the secretary, Nancy Braxton. They spared no efforts to ensure that we arrived safely in Jayapura, enjoyed pleasant accommodations, and had free and easy access to the manuscripts. All of us deeply regret the decision of the Hinduja Foundation to withdraw its support from the Center, which was thriving under the guidance of these two remarkable individuals.

Finally, we must also thank Virendra Nath Sharma of the Fox Valley Campus of the University of Wisconsin, who paved the way for our reception in

Jayapura and photographed the instruments at the Observatory there in the months preceding our visit. We hope that some way will be found to make these photographs available to intersted scholars. And we owe a debt of gratitude to Setsuro Ikeyama, who patiently and accurately entered this complete catalogue into the computer.

David Pingree
Providence, RI

Abbreviations

ASS:	*Ānandāśrama Sanskrit Series*
CESS:	D. Pingree, *Census of the Exact Sciences in Sanskrit,* Series A, vols. 1–6, Philadelphia 1970–
CSS:	*Chowkhamba Sanskrit Series*
IJHS:	*Indian Journal of the History of Science*
IPTS:	*Islamic Philosophy Theology and Science*
KSS:	*Kāśī Sanskrit Series*
PAPS:	*Proceedings of the American Philosophical Society*
RORI:	Rājasthān Oriental Research Institute (Jodhpur)
RPG:	*Rājasthāna Purātana Granthamālā*
SATE:	D. Pingree, *Sanskrit Astronomical Tables in England,* Madras 1973
SATIUS:	D. Pingree, *Sanskrit Astronomical Tables in the United States,* Philadelphia 1968
SBG:	*Sarasvatī Bhavana Granthamālā*

INTRODUCTION

The manuscripts belonging to the library of the Maharaja Sawai Man Singh II Museum at Jayapura are divided into four separate collections: the Khāsmohor ("personal seal") collection, which was originally the personal library of the Mahārājas; the Puṇḍarīka, which was the collection belonging to the Puṇḍarīka or Pauṇḍarīka family descended from Ratnākara Pauṇḍarīka, who was the officiating guru at many of the performances of Vedic sacrifices patronized by Jayasiṃha; the Pothikhānā, which consists of manuscripts copied by the Pothikhānā staff; and the Museum collection, many of the manuscripts of which are on display in the Museum. Except for the Puṇḍarīka collection, these manuscripts were collected by various of the Mahārājas, first at Amber from the reign of Bhāramalla (1547 to 1573) till Jayasiṃha's foundation of Jayapura in 1727,[1] and thereafter at Jayapura itself until the reign of Sawāī Mānasiṃha II (1922 to 1970).[2]

I. The Khāsmohor collection contains at present manuscripts identified by a sequence of nearly 8,000 numbers (many numbers cover several manuscripts); an index of titles and of authors (divided into Sanskrit and vernacular texts) was published in 1976 by the scholar who has done most to catalogue the Jaipur manuscripts, Gopal Narayan Bahura.[3] This has been of immense value to us as the Khāsmohor collection contains 205 manuscripts on astronomy (not counting the pañcāṅgas or calendars, which we have not included in our catalogue). All but eleven of these lie between the library numbers 4951 and 5609, reflecting an attempt to group manuscripts by subject matter. The latest dated manuscript among these is 5122(b) (our catalogue 89), which was copied in 1788/9. This date falls during the reign of Sawāī Pratāpasiṃha (1778 to 1803), who was well known for his interest in the manuscript library that he had inherited and worked to increase.[4]

According to James Tod,[5] Pratāpasiṃha's son, Sawāī Jagatsiṃha (1803 to 1818), whom he describes as "the most dissolute of his race," was so devoted to his concubine Raskafur that: "In the height of his passion for this Islamite concubine, he formally installed her as queen of half his dominions, and actually conveyed to her in gift a moiety of the personality of the crown, even to the invaluable library of the illustrious Jai Singh, which was despoiled, and its treasures distributed amongst her base relations." Bahura[6] notes that the lady probably would not have appreciated manuscripts and would undoubtedly have preferred other objects to satisfy "her lust for wealth".

Inventories of the contents of the library were periodically made. They are now preserved in the Rajasthan State Archives at Bikaner. Unfortunately, these inventories have not as yet been investigated in order that the history of the fate of individual manuscripts in the library could be traced. The astronomical and astrological books listed in one such inventory have been published by Sharma,[7] who claims that they were in Jayasiṃha's personal library. I list here those among these which are astronomical; the ones followed by a number are included in our catalogue of the Khāsmohor collection (we omit those copied after 1743, the year of Jayasiṃha's death). The numbers in parentheses following the titles refer to the number of copies.

Āryabhaṭīya, golādhyāya (1)	
Āryasiddhānta (1)	
Bṛhaspatisiddhānta (1)	
Bhāsvatī (3)	49 (1606/7); 50 (1699); 51 (1701); 52 (1706); 54
Brahmatulyasāraṇī (1)	100–103
Candrasūryagrahaṇādhikāra (1)	
Grahalāghava (6)	72 (1647); 73 (1653); 74 (1704); 75 (1705); 76 (1706); 77(1733); 78 (1741); 81–85
Grahalāghavaṭīkā (1)	76 (1706), 84
Grahalāghavodāharaṇa (1)	82, 86 (1667), 87 (1706), 88 (1710), 91, 92
Karaṇakutūhala (1)	55 (1572); 56 (1705); 57 (1711); 61–63
Karaṇaprakāśikā (1)	47 (1626/7)
Mahādevīsāraṇī (1)	108–110, 112
Makaranda (1)	113–114
Makarandaviveka (1)	116(?)
Mitākṣarā ṭīkā (1) (see *Vāsanābhāṣya*)	
Nalikābandhanakarmapaddhati (1)	244 (1706)
Pātasāraṇī (3)	121; *cf.* 122 (1705)
Pitāmahasiddhānta (1)	3
Rājamṛgāṅka (1)	46 (1618)
Siddhāntakaustubha majastī (1)	
Siddhāntarahasyaṭīkā (1) (see *Grahalāghavaṭīkā*)	
Siddhāntasaṃhitāsāra (1)	247
Siddhāntasindhu Nityānandī (2)	266; 267 (1727); 268; 269
Siddhāntaśiromaṇi (6)	22 (1652); 23 (1651); 24 (1657); 25 (1706); 26 (1709); 27 (1727); 28 (1733); 29–31
Siddhāntaviveka (1)	42; 43
Śrīyantracintāmaṇi (1)	238 (1706)
Sudhāsāraṇī (1)	124(?)
Sundarasiddhānta (1)	41 (1706)

Surasiddhānta (1)	(?)
Sūryasiddhānta (7)	6 (1579); 7 (1642); 8 (1705); 11 (1707); 12; 13; 16
Sūryasiddhānta with ṭīkā (1)	9–10 (1706); 14; 15
Sūryatulyavidhāna (1)	71
Tattvaviveka (1) (see *Siddhāntaviveka*)	
Tithicintāmaṇisāraṇī (1)	123, 180
Tithivinodasāraṇī (1)	
Tripraśnādhikāra (1)	
Vāsanābhāṣya pātādhyāya (1)	26 (1709)
goládhyāya (1)	23 (1651); 25 (1706); 28 (1733); 31; 33
Vāsanābhāṣyamitākṣarā (1)	24 (1657); 26 (1709); 32
Vāsanāyantrādhyāya (1)	25 (1706)
Vyāsasiddhānta (1)	37
Yantrarāja (5)	229 (1596); 230 (1617); 231 (1617); 232 (1706); 233 (1706)
Yantrarājaṭīkā (1)	229 (1596); 230 (1617); 231 (1617); 232 (1706); 233 (1706)
⟨*Zīj-i*⟩ *Muḥammad Shāhī* in Nāgarī (1)	

Just six manuscripts from the above list are definitely not now in the Khāsmohor collection; but the list does not include many manuscripts that we know were in Jayasiṃha's library, some of which still are at Jayapura. The manuscripts definitely in that library include those that he had copied by Tulārāma and Nātha when he first began to study astronomy in about 1706.[8]

9. *Sūryasiddhānta* with bhāṣya of Caṇḍeśvara. Nātha. 6 Nov. 1706.
10. Idem with idem. Nātha. 30 Nov. 1706.
11. *Sūryasiddhānta*. Tulārāma. 1 March 1707.
21. *Sauravāsanā*. Nātha. 31 August 1706.
26. *Siddhāntaśiromaṇi* with *Vāsanābhāṣya*. Tulārāma. 8 Sept. 1709.
41. *Sundarasiddhānta*. Tulārāma. 5 Nov. 1706.
52. *Bhāsvatī*. Tulārāma. 8 Dec. 1706.
76. *Grahalāghava* with ṭīkā. Tulārāma. 5 Sept. 1706.
87. *Grahalāghavodāharaṇa*. Tulārāma. 11 Oct.–26 Oct. 1706.
118. *Makarandodāhṛti*. Tulārāma. 8 Dec. 1706.
232. *Yantrarājāgama* with vyākhyā. Tulārāma. 23 Sept. 1706.

233. Idem. Tulārāma. 16 Nov. 1706.

238. *Yantracintāmaṇi*. Tulārāma. 14 Nov. 1706.

244. *Nalikābandhakarmapaddhati*. Tulārāma. 17 Sept. 1706.

One further manuscript from this group has gone astray. This is a copy of Kamalākara's *Śeṣavāsanā*. 42ff. Copied by Tulārāma at the command of Mahārājādhirāja Jayasiṃha on 1 kṛṣṇapakṣa of Kārttika in Saṃ. 1765 = ca. 18 September 1708. This is now 417 of 1884/86 at the Bhandarkar Oriental Research Institute, Puṇe. Of these 15 manuscripts only the *Śeṣavāsanā* and the *Sauravāsanā* seem not to be included in the inventory.

Another group that we know Jayasiṃha had were manuscripts based on Islamic astronomy that he acquired in the late 1720s; they include copies of translations from Arabic and Persian that are uniformly bound.[9]

253. *Ukara*. ⟨Lakṣmīdhara⟩. 23 October 1729; acquired 1730.

255. *Jarakālīyantra*.

257. *Yantrarājasya rasāla*. Kṛpārāma. ca. 1729.

259. *Śarahatajkirā Virjandī*. Kṛpārāma; acquired 1730.

260. *Vakramārgavicāra*. Lakṣmīdhara.

262. Lunar tables of the *Ulakabegījīca*.

265. *Hayatagrantha*. ⟨Tīkārāma⟩. 1728/9.

267. *Siddhāntasindhu*. Gaṅgārāma. 1727.

270. *Yantraprakāra*.

We also know that Jayasiṃha had a copy of Nityānanda's *Sarvasiddhāntarāja*, since he quotes from it.[10] Of these texts only the name of the *Siddhāntasindhu* is found in the inventory.

Yet another group of manuscripts that should have been in Jayasiṃha's library are those of works that he himself or his associates wrote. Jayasiṃha's works on astronomy are the *Sūryasiddhāntasāravicāra* of ca. 1715/20 (manuscript 44), the *Yantrarājaracanā* of ca. 1720 (not in the Khāsmohor collection, but there is a copy, manuscript 240, in the Puṇḍarīka collection, and a copy was made at the court in Jayapura and presented to David Eugene Smith at Jayapura in 1907; it is now Smith Indic 73 at Columbia University), the *Yantraprakāra* of 1729 (manuscript 270, now in the Museum collection, was presumably once in the Khāsmohor; manuscript 271 is in the Puṇḍarīka collection), and the *Jayavinodasāriṇī* of 1735 (there are no manuscripts left at Jayapura, though pañcāṅgas following its method have been produced there up to the present day. One of the five extant manuscripts—11839 in the Rājasthan Oriental Research Institute in Jodhpur—was copied by Karuṇākara Pau⟨ṇḍarīka⟩ on 28 September

1906[11]). Only two of Jayasiṃha's own four productions, then, are still in the Khāsmohor collection, and none are included in the inventory.

Jayasiṃha's Jyotiṣarāya, Kevalarāma, an astronomer from Modasa in Gujarāt who entered his service in 1725, wrote six works at the Mahārāja's command: the *Bhāgavatajyautiṣayor bhūgolakhagolavirodhaparihāra*, the *Brahmapakṣanirāsa*, the *Dṛkpakṣasāraṇī* in about 1732, the *Vibhāgasāraṇī*, the *Jīvāchāyāsāriṇī*, and the *Pañcāṅgasāriṇī* in 1735. None of these works is represented in the Khāsmohor collection or included in the inventory, but there are two manuscripts—171 and 172—of the *Pañcāṅgasāriṇī* in the Puṇḍarīka collection.

Jayasiṃha's chief guru, Jagannātha Samrāṭ, translated Euclid's *Elements* and Ptolemy's *Almagest* from Arabic into Sanskrit. One manuscript containing both these translations is now number 32 of the Reserved Collection of the Museum collection; we were not allowed to examine it because of pending litigation. It presumably was once in the Khāsmohor collection, but is not included in the inventory. A second copy, perhaps of the Reserved Collection manuscript, is in the Pothikhānā collection (254). There were three versions of Jagannātha's *Siddhāntakaustubha*; the first, written in early 1727, is in the Puṇḍarīka collection (45);[12] the last, presented as a supplement to the translation of the *Almagest* in 1732, is apparently the *Siddhāntakaustubha majastī* of the inventory. At least part of Jagannātha's library was acquired by Viśveśvara Pauṇḍarīka, whose floruit was toward the end of the eighteenth centry. His acquisitions included manuscripts 36 (*Vṛddhavasiṣṭhasiddhānta*. Copied 10 October 1690), 39 (*Somasiddhānta*. Copied 13 October 1723), and 123 (*Subodhinī*. Copied 15 November 1730), but apparently none of Jagannātha's own works.

It is quite surprising that so many of the treatises produced by these three men at Jayasiṃha's court are no longer in the Jayapura library, but it is even more astonishing that only one is included in an inventory that allegedly comes from Jayasiṃha's own time, and that that is a work written in 1732.

The last item in the inventory is ⟨*Zīj-i*⟩ *Muḥammad Shāhī*, allegedly in Nāgarī. There are several manuscripts at Jayapura with material from Ulugh Beg's *Zīj-i Jadīd*; these are manuscripts 261–264. Ulugh Beg's planetary tables and lunar tables (261 and 262) are in the Museum and Khāsmohor collections; the star catalogues, precessed to 1726, are in the Puṇḍarīka collection. Of course the precessed star-catalogue of Ulugh Beg was incorporated into the Persian *Zīj i Muḥammad Shāhī*, but it is doubtful that this is what the inventory refers to, especially as it is not in the Khāsmohor collection. A more likely candidate is a manuscript now in Berlin, or. fol. 2973 at the Städtbibliothek.[13] This contains computations of the longitudes and latitudes of the Moon and the planets and of the longitudes of the Sun at noon at Jayapura on 7 March 1718 according to the *Zīj i Muḥammad Shāhī* and computations according to the same authority of a lunar eclipse on 28 May 1732 and of a solar eclipse visible at Jayapura on 20 May 1734. This manuscript may well have once been in the Khāsmohor collection. I do not know how it got to Berlin.

The inventory, then, bears some relationship to the Jayapura library. If the reference to the *Zīj i Muḥammad Shāhī* is part of the original list, that list was made after about 1735, otherwise after 1732. But it was clearly not a complete inventory, whenever it was compiled.

II. The Puṇḍarīka collection, which consists of about 2,300 manuscripts collected by Ratnākara Pauṇḍarīka and his descendents between the early eighteenth and the late nineteenth century, was acquired by the Jayapura Pothikhānā under Sawāī Mādhavasiṃha II in 1905[14] and added to the existing library. We were fortunate to be given access to a handwritten, classified list, from which we identified fifty-seven manuscripts to include in this catalogue. Ratnākara belonged to a Mahārāṣṭrian family of the Śauṇḍilyagotra, named Mahāśabde. His father, Devabhaṭṭa, who had been honored by Jayasiṃha's great-grandfather, Mahārāja Rāmasiṃha (1667 to 1689),[15] was a resident of Kāśī, where his son, Ratnākara, studied under the famous scholar Nāgeśa (Nāgojī) Bhaṭṭa. He was summoned by Jayasiṃha to Amber to be his guru shortly after his ascent of the *gaddi* of Amber on 25 January 1700; for Ratnākara at the conclusion of his *Jayasiṃhakalpadruma*, which he completed on ca. 21 August 1713, reports that he had already performed for Jayasiṃha the following Vedic sacrifices: *jyotiṣṭoma*, *vājapeya*, and *pauṇḍarīka* (for which he received the title Pauṇḍarīkayājin).[16] It is known that he performed the *vājapeya* for Jayasiṃha on 31 December 1708.[17]

Nothing further is known about Ratnākara, who must have been quite elderly in 1713; he probably died about 1720. His son Sudhākara performed a *puruṣamedha* sacrifice for Jayasiṃha at some uncertain time.[18] Another son, Gaṅgārāma, who had participated in the *vājapeya* sacrifice in 1708,[19] and to whom Jayasiṃha paid his respects in 1733, died in 1755.[20]

The members of the family who are known to have acquired manuscripts on astronomy lived in the latter half of the eighteenth century. The most important was Viśveśvara, the son of Rāmeśvara[21], the son of Gaṅgārāma, the son of Ratnākara, whose acquisition of at least three of Jagannātha's manuscripts has already been reported. The other manuscripts that he owned included 48 (*Karaṇaprakāśa*), 60 (*Karaṇakutūhala*. Copied by Gopīnātha, the son of Rāmeśvara Vyāsa, on 31 October 1788), 65 (*Brahmatulyodāharaṇa*), 90 (*Grahalāghavodāharaṇa*. Copied in 1788/9), 212 (*Chādikanirṇaya*), 236 (*Dhruvabhramayantra*. Copied by Gopīnātha, the son of Rāmeśvara Vyāsa, on 5 November 1789), 240 (*Yantrarājaracanā*), and 258 (*Yantrarājasya rasāla*. Copied for the Pauṇḍarīkas on 28 November 1788). Viśveśvara Mahāśabda Pauṇḍarīka also composed 243 (*Palabhāyantra*), probably in the 1790s; he was the author of the tables in manuscript 199 as well. Viśveśvara also wrote a *Nirṇayakautuka* and a *Pratāpārka* on dharmaśāstra for Mahārāja Pratāpasiṃha.

Another member of the Pauṇḍarīka family, Gokula or Gokulanātha, the son of Śambhūnātha, copied 93 (*Grahalāghavodāharaṇa*), 171 and 172 (*Pañcāṅgasāraṇī*. Copied 3 April and 28 April 1793), and 189 (a sāyanalagnasāriṇī. Copied in November or December 1788). He also employed Gopīnātha—presumably the son of Rāmeśvara Vyāsa, who had worked for Viśveśvara—to copy 152 (*Grahalāghavasāriṇī*). Gopīnātha, the son of Rāmeśvara Vyāsa, also copied 234 (*Yantrarājāgama*) for Jāgeśvara on 14 December 1788; it is not known whether or not Jāgeśvara was a Pauṇḍarīka.

The last Pauṇḍarīka whom we know to have contributed manuscripts on astronomy to the family collection was Dhaneśvara who owned 225 (a treatise on stars) and 228 (*Palabhājñāna*) and who employed Śyāmasundara to copy 80 (*Grahalāghava*) on 6 March 1826. It was probably the same scribe who copied the same text on 16 January 1826 (manuscript 79), and who seems to have been involved in 245 (on the gnomon).

Certain other manuscripts in the Puṇḍarīka collection clearly represent or are manuscripts that were once in the Khāsmohor collection. These are 48 (*Karaṇaprakāśa*. A copy of Khāsmohor 4953), 94 (*Grahalāghavavārttika*. A commentary that illustrates the astronomy of the *Grahalāghava* by computing according to it the longitudes of the planets at the time of Jayasiṃha's birth on 3 November 1688), 222–223 (sūryapattras), 234 (*Yantrarājāgama*. Copied from Khāsmohor 4958 by Gopīnātha on 14 December 1788), 246 (on gnomonics as practiced at Jayapura), and 256 (*Jarakālīyantra*).

III. The Pothikhānā collection, which was instituted by Jayasiṃha, comprises copies of manuscripts in the Khāsmohor collection, copies of other manuscripts and printed books, and manuscripts presented as gifts. There are about 2,200 manuscripts in this collection: catalogues have been published of its holdings on *dharmaśāstra*[22] and *stotras*.[23] We consulted a handwritten list of the jyotiṣa manuscripts, and found just five that fit the criteria for inclusion in this catalogue. Four of these were copied by the same scribe from printed books; these are 4 (*Āryabhaṭīya*), 5 (*Śiṣyadhīvṛddhidatantra*. Copied by Līchamīnārāyaṇa at Surata on 10 July 1895), 17 (*Sūryasiddhānta*), and 35 (*Brahmasiddhānta*. Copied by Līchamīnārāyaṇa at Surata on 24 August 1895). The last manuscript is 254 (*Rekhāgaṇita* complete and *Samrāṭsiddhānta* incomplete), perhaps a copy of manuscript 32 of the Reserved Collection.

IV. The Museum collection presumably contains manuscripts extracted from the Khāsmohor collection as being especially rare and valuable, and suitable for public display in the Museum. A list of these manuscripts has been published;[24] we have included twelve in our catalogue. We did not have sufficient time to include the fourteen manuscripts of astronomical works in Arabic and Persian.[25] We hope to be able to remedy this failure on a later occasion.

In conclusion, we should comment on some of the most valuable manuscripts that we discovered in cataloguing this collection. We will discuss those in each category in its order in the catalogue.

B. Siddhāntas

3. The *Paitāmahasiddhānta* from the *Viṣṇudharmottarapurāṇa*. One of four copies of this work, fundamental to the Brahmapakṣa. David Pingree hopes to prepare a new edition.

44. The *Sūryasiddhāntasāravicāra* attributed to Jayasiṃha himself; there is only one other known copy. The Jayapura copy is important especially for its diagrams. An edition is being prepared by Gary Tubb.

45. The earliest version of Jagannātha's *Siddhāntakaustubha*, in which Jayasiṃha's bijas to the *Sūryasiddhānta* devised in 1726 and his parameters derived

from Nityānanda's *Sarvasiddhāntarāja* are recorded. David Pingree has prepared an edition of this.

C. Karaṇas

46. A full version of Bhojadeva's *Rājamṛgāṅka* radically different from the published version. David Pingree will publish this.

71. Dāmodara's *Sūryatulya*. Only three other copies are known.

D. Koṣṭhakas

118 The unique manuscript of Moreśvara's *Makarandaṭippaṇa*.

143. The unique manuscript of Harinātha's tithi, nakṣatra, and yoga tables.

147. The unique manuscript of Goparāja's *Khagataraṅgiṇī*.

171 and 172. Two of the four known manuscripts of Kevalarāma's *Pañcāṅgasāriṇī*. An edition has been prepared by David Pingree.

E. Eclipses

216–220. Observations and computations of lunar eclipses on 18 May 1761, 17 April 1772, 23 March 1773, 30 September 1773, 15 February 1775, 5 February 1860, and 6 February 1860 (sic!).

F. Star Charts

221. The unique star chart of Mādhavasiṃha, made in about 1760.

222 and 223. Unique sūryapattras.

G. Geographical Tables

226 and 228. Two geographical tables supplementing those previously published.[26]

I. Miscellaneous

250–251. Two manuscripts of Nandarāma Miśra's *Bhāgavatajyotiḥśāstrabhūgolakhagolavirodhaparihāra*, an expansion of Kevalarāma's similarly entitled treatise. Editions of both are being prepared by Christopher Minkowski and David Pingree.

J. Translations

259. The unique copy of the Sanskrit translation of II 11 of Naṣīr al-Dīn al-Ṭūsī's *Tadhkira* with al-Birjandī's commentary. An edition has been published by Takanori Kusuba and David Pingree.

263 and 264. Sanskrit versions of Ulugh Beg's star catalogue. These will be edited by David Pingree.

265. A rare manuscript of the Sanskrit translation of 'Alī al-Qūshjī's *Risālah dar hay'at*. An edition by David Pingree and Kim Plofker is in preparation.

266–269. Four of the six manuscript copies of Nityānanda's Sanskrit translation of Farīd al-Dīn Dihlawī's *Zīj i Shāhjahānī*.

271. The third known copy of Jayasiṃha's *Yantraprakāra*, which has been published from the other two by S. R. Sarma.

272. The manuscript copy of the second edition of Philippe de La Hire's *Tabulae Astronomicae*, published at Paris in 1727, completed by Joseph du Bois at Jayapura on 10 September 1732.

273 and 274. Two of the four manuscripts containing the prose Sanskrit translation of de La Hire's *Tabulae* and/or the *Phiraṅgicandracchedyopayogika*. Both texts will be published by David Pingree.

275. Attempts to illustrate de La Hire's third lunar equation.

276. Tables comparing Jayasiṃha's observed lunar positions with those computed by means of de La Hire's tables and the "New Tables."

The thorough investigation of these manuscripts will greatly expand our knowledge of the astronomical activities at Jayasiṃha's court, which constitute in many aspects a sort of rehearsal for the Indian reaction to Western astronomy in the nineteenth century.

The Manuscript Descriptions

Each separate text is headed by a paragraph giving its title or titles, its author if known, the place of its composition if known, and its date if known. These data are followed by references to *CESS* (D. Pingree, *Census of the Exact Sciences in Sanskrit*, Series A, vols. 1–5, Memoirs of the American Philosophical Society, Philadelphia: American Philosophical Society, 1970–1994), *SATIUS* (D. Pingree, *Sanskrit Astronomical Tables in the United States*, Transactions of the American Philosophical Society 58,3, Philadelphia: American Philosophical Society, 1968), and *SATE* (D. Pingree, *Sanskrit Astronomical Tables in England*, Madras, Kuppuswami Sastri Research Institute, 1973), as appropriate, and a reference to the edition, if one exists, with which the manuscript has been compared. After this is given, whenever available, the work's incipit—usually a verse.

The description of the manuscripts of each work are arranged first in ascending order of the dates of copying for those that are dated, then in the ascending order of their identification numbers in the collection they belong to.

The description of each manuscript consists of its serial number, its shelfmark, the numbers given by the scribe to its folios or, lacking those, an assignment of capital English letters to unnumbered folios, the dimensions of the manuscript leaves in centimeters to the nearest half-centimeter, height before width (the writing is always taken to be parallel to the width), and the number of lines per page. After this, as appropriate, are comments on the physical condition of the manuscript, the presence of marginalia, the part of the text included, and the presence of a commentary. Then, as available, is information about the scribe, his location, and the date on which he completed the copying of the manuscript. Finally, any information concerning owners of the manuscript is recorded.

If the manuscript is incomplete, the beginning and end of each portion of the text (but not the incipit if nothing is missing from the beginning of the text) is given with the page and line in the edition referred to on which the first and last words occur. Descriptions of koṣṭhakas generally give details concerning the structure, use, and parameters of each table.

Succeeding this is the colophon with a reference to the folio on which it occurs; the colophon is the statement of the author and the title of the text.

After this is the post-colophon, which gives information concerning the scribe, his patron, and the place and date of the copying. Ownership notes, if any, follow this. These excerpts from the manuscript are presented in the scribe's orthography.

Next are given copies of material extraneous to the main text that the original scribe or later persons have written in blank spaces in the manuscript—usually on the recto of the first leaf and/or the verso of the last. The final element is an indication of the price of the manuscript.

The volume ends with a concordance of library numbers and catalogue numbers of the manuscripts, a list of the dates on which manuscripts were copied, and indices of authors, titles, scribes and owners, other owners, families and other social units, toponyms, and subjects. The reference numbers are the catalogue numbers of the manuscripts.

David Pingree

Notes

[1]Gopal Narayan Bahura, *Literary Heritage of the Rulers of Amber and Jaipur*, Jaipur 1976, first part, pp. 27–49.

[2]*Ibid.*, pp. 71–95.

[3]*Ibid.*, second part, pp. 3–130 (Sanskrit works), pp. 131–217 (bhāṣā works), and p. 218 (Arabic, Persian, and Urdu works); pp. 219–287 (Sanskrit authors), pp. 288–342 (bhāṣā authors), and pp. 343–500 (extracts from important manuscripts).

[4]*Ibid.*, first part, pp. 17–18 and 80–81.

[5]James Tod, *Annals and Antiquities of Rajasthan*, edited by William Crooke, 3 vols., London: Oxford University Press, 1920, vol. 3, pp. 1364–1365.

[6]Bahura, *op. cit.*, first part, p. 18.

[7]Virendra Nath Sharma, *Sawai Jai Singh and His Astronomy*. Delhi: Motilal Banarsidass, 1995, pp. 328–330.

[8]David Pingree, "An Astronomer's Progress," *Proceedings of the American Philosophical Society* 143, 1999, 73–85. I have reinterpreted the ambiguous wordings of the dates of manuscripts 9, 26, 41, 232, 233, 238, and 244 in an attempt to improve the data given on p. 74, fn. 7.

[9]D. Pingree, "Sanskrit Translations of Arabic and Persian Astronomical Texts at the Court of Jayasiṃha of Jayapura," *Suhayl* 1, 2000, 101–106.

[10]D. Pingree, "The Original Version of Jagannātha's *Siddhāntakaustubha*," forthcoming.

[11]The manuscripts and the tables of Jayasiṃha's *Jayavinodhasāriṇī* are described by D. Pingree, "Kevalarāma's *Pañcāṅgasāriṇī*", forthcoming.

[12]For these versions and their interrelation see the article cited in fn. 10.

[13]D. Pingree, "Indian Reception of Muslim Versions of Ptolemaic Astronomy," in F. Jamil and Sally P. Ragep, eds., *Tradition, Transmission, Transformation*, Leiden: E. J. Brill, 1996, 471–485 (484).

[14]Bahura, *op. cit.*, first part, pp. 21 and 89.

[15]*Ibid.*, first part, p. 44, and second part, p. 424.

[16]*Jayasiṃhakalpadruma*, edited by Harinārāyaṇa Śarman, Kalyāṇa-Muṃbaī, 1925, p. 912.

[17]Bahura, *Literary Heritage*, first part, p. 59, and second part, p. 423. This yajña is described by Viśvanātha Bhaṭṭa in his *Rāmavilāsakāvya* III 1–51, ed. G. N. Bahura, Jaipur: Maharaja Sawai Mansingh II Museum, 1978, pp. 46–53.

[18]V. S. Bhatnagar, *Life and Times of Sawai Jai Singh*, Delhi: Impex India, 1974, p. 344.

[19]Bahura, *op. cit.*, second part, p. 425.

[20]Bhatnagar, *op. cit.*, p. 332; see also K. V. Sarma "Brahmapuri, the Scholars' Village Established by Maharaja Sawai Jai Singh," *Indologica Jaipurensia* 2, 1988–1995, 45–53, esp. 48–49.

[21]On Rāmeśvara see Sarma, *op. cit.*, 49-50.

[22]G. N. Bahura, *Catalogue of Manuscripts in the Maharaja Sawai Man Singh II Museum (Pothikhana Collection; (a) Dharmaśāstra)*, Jaipur: Maharaja Sawai Man Singh II Museum, 1984.

[23]G. N. Bahura and R. G. Sharma, *Catalogue of Manuscripts in the Maharaja Sawai Man Singh II Museum (Pothikhana Collection; (b) Stotras)*, Jaipur: Maharaja Sawai Man Singh II Museum, 1987.

[24]G. N. Bahura, *Catalogue of Manuscripts in the Maharaja of Jaipur Museum*, Jaipur: Maharaja of Jaipur Museum, 1971, pp. 2–63 (Sanskrit), pp. 64–69 (Hindī), pp. 70–77 (Arabic and Persian), pp. 76–77 (Latin), p. 78 (Reserved Collection), and pp. 79–126 (details of important manuscripts).

[25]*Ibid.*, pp. 72–77; David A. King, "A Handlist of the Arabic and Persian Astronomical Manuscripts in the Maharaja Mansingh II Library in Jaipur," *Journal for the History of Arabic Science* 4, 1980, 81–86; and D. Pingree, "Indian and Islamic Astronomy at Jayasiṃha's Court," in D. A. King and G. Saliba, eds., *From Deferent to Equant*, New York: New York Academy of Sciences, 1987, pp. 313–328 (pp. 313 and 326, fn. 4).

[26]D. Pingree, "Sanskrit Geographical Tables," *IJHS* 31, 1996, 173–220.

CATALOGUE
A. VEDIC

The **JYOTIṢAVEDĀṄGA** in the Ṛk recension composed by Lagadha in Northwest India (Gandhāra?) in *ca.* -400. *CESS* A5, 538a–539a. Edited by S. Dvivedin, Benares 1908, pp. 61–69. The first verse is:

pañcasaṃvatsaramaya-
yugādhyakṣaṃ prajāpatim |
dinartvayanamāsāṅgaṃ
praṇamya śirasā śuciḥ ||

Manuscripts:

1. Puṇḍarīka veda 262. Ff. 1–3. 11 × 25 cm. 8/9 lines. Copied by Rāmabhaṭa on Monday 9 October 1797, perhaps at Kāśī (at the beginning is: śrīgaṇeśāya namaḥ || śrīkāśīviśveśvarāya namaḥ).

Colophon on f. 3v: iti jyotiṣa samāptaḥ.

Post-colophon: saṃvata 1854 kārtikavadya 4 iṃduvāsare.

On f. 1 is written: idaṃ pustakaṃ rāmabhaṭena likhitaṃ.

2. Puṇḍarīka veda 261. Ff. 1–3. 9.5 × 22.5 cm. 9–11 lines. Some marginal notes. Perhaps copied in Kāśī.

Colophon on f. 3v: iti jyotiṣaḥ samāptaḥ.

After this is written: paṃcasaṃvatsaraṃ prapadyate kāryāḥ kalā daśa ca yāḥ parva savitāviṣuvaṃ sapta || śrīkāśīviśveśvarārpaṇam astu.

After this a second scribe has written:

kāliṃdīpulinīsuradrumavanīvānīrakuṃjasthalīṃ
pāhātāṃkarilājivāsicarilā ghorā kaṭhorā⟨ṃ⟩galīṃ
prālemdī navamī navīnavaṇitāṃ joḍīṃ tulā aṃjalī
sākṣī he rajanī manojasajanī aisā harīhāchalī 1
vegūnī guruśailasarvaśikhare[a] dvārāpagā laṃghunī
prārgī lokapayodadṛṣṭikarakā vṛṣṭiṃ śirīṃ sā unī
cittīṃ darśanalālasādharuniyāṃ ālemṃ aśīsāhaśī
ātāpuṇyavalemṃ tu syā priyatamā ho ūnalokīṃ haśī 2

a. °śi$\overset{2}{\text{r}}\overset{1}{\text{e}}$kha

B. SIDDHĀNTAS

The **PAITĀMAHASIDDHĀNTA** presented as a conversation between Bhṛgu and Bhagavān (Brahmā) and preserved in the *Viṣṇudharmottarapurāṇa* as adhyāyas 168–174 of Khaṇḍa 2; it was composed in about 425. See *CESS* A4, 259a. Edited by V. Dvivedin in *Jyautiṣasiddhāntasaṅgraha*, Benares 1912, fasc. 2, pt. 1. It begins: puṣkara uvāca | atha bhagavantaṃ bhuvanotpattisthitisaṃhārakārakaṃ carācaragurum pratiyaśasaṃ samadhigamya bhṛgur vijñāpayāmāsa.

Manuscript:

3. Museum 43. Ff. 1–14. 12 × 26 cm. 9 lines.
Colophon on f. 14: iti viṣṇudharmottareṣu paitāmahasiddhāṃtaḥ samāptaḥ
Price on f. 14v: kīmati | ≡.

The **ĀRYABHAṬĪYA** composed by Āryabhaṭa at Kusumapura = Pāṭaliputra in *ca.* 500. *CESS* A1, 50b–53b; A2, 15b; A3, 16a; A4, 27b; and A5, 16a–17a. Edited with the ṭīkā, *Bhaṭadīpikā*, of Parameśvara by H. Kern, Leiden 1874. The first verse is:

praṇipatyaikam anekaṃ
kaṃ satyāṃ devatāṃ paraṃ brahma |
āryabhaṭas trīṇi gadati
gaṇitaṃ kālakriyāṃ golam ||

Manuscript:

4. Pothikhānā 22. Ff. 1–41 and 9ff, of which the first 4 are numbered 1, 2, 4, and 5 (text continuous). 30.5 × 19.5 cm. European style book bound in cardboard covered with blue cloth having a flower design; hinged to left. Copied by the scribe of manuscripts 5, 17, and 35.

Ff. 1–41. The *Āryabhaṭīya* with the *Bhaṭadīpikā* of Parameśvara.

Colophon on f. 41: iti bhaṭadīpikāyāṃ golapādaḥ ity āryabhaṭīyaṃ samāptaṃ.

After this is written: śāstroditai++++vāṃśeś ca tāsām udayalagnaṃ madhyalagnam astalagnaṃ[a] ca samyag jñātvā punar arkasyārdhāstamaye ghaṭikāyaṃtraṃ saṃsthāpya tena kṛtikādīnāṃ + dyena kālena viśe + ⁕ pratikañcuko yo 'sya iti pavanīyaṃ dīpikāvyākhyāyā vyākaraṇaviruddhatvāt.

a. astalagnaś

F. 41v. Blank.

Ff. ⟨42–47⟩. The *Āryabhaṭīya* alone except for the last verse of the *Bhaṭadīpikā* on f. ⟨47⟩:

paramādīśvarākhyena kṛteyaṃ bhaṭadīpikā
pradīyyatāṃ sadā jyotiśśāstrajñānāṃ hṛdālaye ||

Colophon on f. ⟨47⟩: iti bhaṭadīpikāyāṃ golapādaḥ ity āryabhaṭīyaṃ samāptaṃ.

Ff. ⟨47v–50v⟩. Blank.

The **BHAṬADĪPIKĀ**, a commentary on Āryabhaṭa's *Āryabhaṭīya*, composed by Parameśvara at Aśvatthagrāma (= Ālattūr) in Kerala in *ca.* 1425. *CESS* A4, 189b–190a, and A5, 211b–212a. Edited by H. Kern, Leiden, 1874. The first verse is:

> yattejaḥ prerayet prajñāṃ sarvasya śaśibhūṣaṇam |
> mṛgaṭaṅkābhayeṣṭāṅkaṃ trinetraṃ tam upāsmahe ||

See manuscript 4.

The **ŚIṢYADHĪVṚDDHIDATANTRA** composed by Lalla in Lāṭadeśa in the eighth or early ninth century. *CESS* A5, 545b–546b. Edited by S. Dvivedin, Kāśī 1886. The first verse is:

> natvā brahmaharitrinetradinakṛcchītāṃśubhūnandana-
> prāleyāṃśusutendramantribhṛgujacchāyāsutebhānanān |
> ācāryāryabhaṭoditaṃ suviṣamaṃ vyomaukasāṃ karma yac
> chiṣyāṇām abhidhīyate tad adhunā lallena dhīvṛddhidam ||

Manuscript:

5. Pothikhānā 23. Ff. A–B; ff. 1–25 and ⟨26⟩; f. C; ff. 27–30, 30b, 31–39; and ff. D–E. 30.5 × 19.5 cm. 26 lines. European style book bound in cardboard covered with blue cloth having a flower design; book hinged to left. Copied by Līchamīnārāyaṇa at Surata on Wednesday 10 July 1895; he also copied manuscripts 4, 17, and 35. With Dvivedin's footnotes.

Ff. A–Av. Table of contents.

Ff.B–Bv. Blank.

Ff. 1–⟨26⟩. Grahagaṇitādhyāya.

Ff. 26v and C–Cv. Blank.

Ff. 27–39. Golādhyāya.

Colophon on f. 39: iti śrīlallācāryaviracite śiṣyadhīvṛddhide mahātantre gole praśnādhyāyaḥ samāptaḥ || samāpto 'yaṃ śiṣyadhīvṛddhidaḥ

Post-colophon: līchamīnārāyaṇa lekhata pothīṣānā surataṣānā kī mī° śā° va° 5 saṃ° 1952 vāra vudhavāra°.

Ff. 39v and D–Ev. Blank.

The **SŪRYASIDDHĀNTA** in the form of dialogue between Maya and Sūrya, composed in *ca.* 800. *CESS* A6. Edited with the ṭīkā, *Gūḍhārthaprakāśaka*, of Raṅganātha by F. Hall, Calcutta 1859. The first verse is:

acintyāvaktarūpāya nirguṇāya guṇātmane |
samastajagadādhāramūrtaye brahmaṇe namaḥ ||

Manuscripts:

6. Khasmohor 5360. Ff. 1–15. 11.5 × 26 cm. 14/15 lines. With some marginal notes. Copied by Sahajasundara, the pupil of Saravaṇa Gaṇi, the pupil of the Mahopādhyāya Mativilāsa of the Maladhāragaccha, for himself at Argalapura (= Agra) on Monday 16 November 1579 during the reign of Pātīsāhi Jalāladīni Akavvara.

Colophon on f.15v: iti śrīsūryasiddhānte mayāsurasaṃvāde prājāpatyaviracite mānādhyāyas trayodaśakaḥ

Post-colophon: saṃvat 1636 barṣe | mārgasiravadi 13 somavāsare śrīmāladhāragacche | mahopādhyāyaśrī6mativilāsa | tacchiṣyaga° śrīsaravaṇa | tacchiṣyaṇa° śrīsahajasuṃdara ātmārthaṃ svayam ālekhi || śrīargalapuramadhye | pātīsāhiśrījalāladīni akavvararājye.

After this is written in the scribe's orthography:

lagnasṛṣṭikaṭigrīva || baddhadṛṣṭir adhonmukhaṃ |
kaṣṭena likhitaṃ śāstraṃ yatnena paripālayet || 1
tailād ṛkṣe jalad ṛkṣe|d ṛkṣe śithilabaṃdhanāt |
mūrṣahaste na dātavyaṃ | evaṃ vada (here it ends).

7. Puṇḍarīka jyotiṣa 2. Ff. 1–31. 11.5 × 27 cm. 10 lines. Some folios stuck together. Copied at Pharatapura on Tuesday 1 November 1642.

Colophon on f. 31v: iti śrīsūryasiddhāṃte mayāsurasaṃvade prajāpatyaviracite manādhyāyaś caturdaśamaḥ ||14||.

Post-colophon: saṃvat 1⟨6⟩99 varṣe mārgasirvadi 5 bhaumavāsareḥ || pharatapuramadhye likhitaṃ.

8. Khasmohor 5403. Ff. 1–28, 28v, and 29–34. 11.5 × 24 cm. 8 lines. Copied by Vallabha Bhaṭṭa at Avantīpurī on Tuesday 3 July 1705.

Colophon on f. 34: iti sūryasiddhānte mānādhyāyaś caturdaśaḥ || samāpto yaṃ graṃthaḥ.

Post-colophon: saṃvat 1762 miti śrāvaṇamāse kṛṣṇapakṣe navamyāṃ bhaumavāsare li. bhaṭṭavallabhena avaṃtīpūryyāṃ.

Price on f. 1: kīmati ||| =.

9. Khasmohor 5139. Ff. 1–17. 13 × 24 cm. 10 lines. With occasional pṛṣṭhamātras. With the *bhāṣya* of Caṇḍeśvara. Copied by Nātha, the son of Viṭṭhala Bhaṭṭa of the Medapāṭhajñāti, a resident of Rūpapura, for Jayasiṃha from AS Bombay 293 I in Gujarāta on Saturday 9 November 1706. Incomplete (end of adhyāya 10 to the end of adhyāya 11).

F. 1v begins at the end of the *bhāṣya* on 10, 15: saṃpūrṇṇaśuklā syāt || cha ||

Colophon of adhyāya 10 follows: iti śrīmaithilavājapeyasomayājiśrīcaṃḍeśvarācāryaviracite sūryasiddhāṃtabhāṣve paurṇamāsī caṃdraśṛṃgonnatir ddaśamo dhyāyaḥ || samāptaḥ.

Colophon on f. 17: iti śrīmaithilavājapeyasomayājiśīcaṃḍeśvarācāryaviracite sūryasiddhāṃtabhāṣye vyatipātavaidhṛtādhyāya ekādaśaḥ samāptaḥ.

Post-colophon: saṃvat 1763 varṣe kārttikamāse suklapakṣe puṇyamāsī śanivāsare gujarātamadhye rūpapurame vāstavyaṃ medapāṭhajñātīyabha° viṭhalasūtanāthena laṣitaṃm idaṃ pustakaṃ || || māhārājādhirāja gau brāṃmaṇa prati ciraṃjīvaḥ[a].

a. after ciraṃ° ji is written, but crossed out.

Price on f. 1: kīmatī | ≡J.

10. Khasmohor 5013. Ff. 1–63, 63b(=64), and 65–72. 13 × 23.5 cm. 10 lines. With the *bhāṣya* of Caṇḍeśvara. Copied by Nātha, the son of Viṭṭhala Bhaṭṭa of the Medapāṭhajñāti, from AS Bombay 293 II at Rūpapura in Gujara on Saturday 30 November 1706. Incomplete (adhyāyas 12–14 numbered ⟨11⟩, 12, and 13).

F. 1v begins in 12, 1a (p. 194, 5): athārkāṃśamahābhūtaṃ.

Colophon on f. 72: iti śrīmaithilavājapeyasomayājiśrīcaṃḍeśvarācāryaviracite sir $\overset{11}{\text{v}}$ vas$\overset{11}{\text{ū}}$ryasiddhāṇtabhāṣye trayodaśo 'dhyāyaḥ samāptaḥ.

Post-colophon: saṃvat 1763 varṣe mārgaśiraṣamāse śuklapakṣe śaṣṭīdine śanivāsarena laṣitaṃ medapāṭhajñātī bha°viṭhalasūtanāthena laṣitaṃm idaṃ pustakaṃ gujaramadhye rūpaparamadhye vāstavyaṃ || || cha || || samāpto yaṃ graṃtha ||

Price on f. 72v: kīmatī ๆ|| ≡ ||J.

11. Khasmohor 5302. Ff. 1–34. 11 × 26.5 cm. 9 lines. Copied by Tulārāma for Mahārājādhirāja Jayasiṃha on Saturday 1 March 1707.

Colophon on f. 34: iti śrīsūryasiddhānte mānādhyāyaś caturdaśaḥ.

Post-colophon: śrīmanmahārājādhirājajīśrījayasiṃhadevajīkasyājñayā li. tulārāmeṇa saṃ 1763 mīti phālguṇaśukla 9 śanau.

12. Khasmohor 5142. Ff. 1–3, 3b, and 4–25. 11.5 × 25 cm. 10–12 lines.

Colophon on f. 25 :iti śrīsūryasiddhānte mānādhyāyaś caturdaśaḥ || 14 || samāptim agat.

13. Khasmohor 5145. Ff. 1–34. 11.5 × 23 cm. 9–11 lines. With marginalia. Ff. 1-11 and 13–34 brown paper; f. 12 white paper. 11.5 × 22.5 cm. 9 lines, supplied by first scribe.

Colophon on f. 23v: iti śrīsūryasiddhāṃte mānādhyāyaś caturdaśaḥ || 14 || samāptā.

14. Khasmohor 5187. Ff. 1–53. 12 × 28 cm. Tripāṭha. With some marginal notes. With the ṭīkā, *Sauravāsanā*, of Kamalākara. Incomplete (adhyāyas 1–11).

Colophon of the mūla on f. 53: iti sūryasiddhāṃte pātādhikāraḥ samāptaḥ.

Colophon of the ṭīkā on f. 53: iti śrīsakalagaṇakasārvabhaumaśrīmannṛsiṃhātmajakamalākaraviraviracitā daśādhikārāṃtaṃ pūrvakhaṃḍa sauravāsanā saṃpūrṇā.

15. Khasmohor 5190. Ff. 1–24 and 26–48 (text continuous). 12.5 × 27 cm. 10 lines. With the ṭīkā, *Gūḍhārthaprakāśaka*, of Raṅganātha. Incomplete (adhyāyas 12–14).

f. 1v begins, after the salutations, with the beginning of the ṭīkā on adhyāya 12 (p. 294, 1): mahādevaṃ.

Colophon on f. 48: iti śrīsakalagaṇakasārvabhaumavallāladaivajñātmajaraṃganāthaviracitaḥ sūryasiddhāṃtagūḍhārthaprakāśakaḥ sapūrṇaḥ.

After this is written: grathasaṃkhyā 15 atra pūrvottarārddhayo mitilitvā grathasaṃkhā sapādapaṃcasahasramitā 1350.

16. Khasmohor 5573. Ff. 1–27 and 1 blank folium. 12 × 31 cm. 9 lines.

Colophon on f. 27v: iti śrīsūryasiddhāṃte mayāsurārkkāṃśasaṃvāde mānādhyāyāḥ || 12 || samāpto yaṃ sūryasiddhāṃtaḥ.

17. Pothikhānā 24. Ff. A–B, ff. 1–29(=pp. 1–58), and ff. C–E. 30.5 × 18.5 cm. 19 lines. European style book bound in cardboard covered with blue cloth having a flower design; hinged to left. Copied by the scribe of manuscripts 4, 5, and 35.

Ff. A–B. Blank.

F. Bv. Index of chapters.

Colophon on f. 29v: iti sūryasiddhānte mānādhyāyaḥ || || samāptaś ca sūryasiddhāntaḥ.

Ff. C–Ev. Blank.

The **SŪRYASIDDHĀNTABHĀṢYA**, a commentary on the *Sūryasiddhānta* composed by Caṇḍeśvara in Mithilā in 1185. *CESS* A3, 40b–41a, and A5, 105b. The first verse is:

namas te paramātmaikarūpāya paramātmane |
svecchāvabhāsitāśeṣadehābhinnāya śambhave ||

See manuscripts 9 and 10.

The **SŪRYASIDDHĀNTAṬĪKĀ** composed by Kṛṣṇa at Kāśī in 1584. *CESS* A5, 48a–48b. The first verse is:

praṇamya jagadādhāraṃ nirādhāraṃ ca nirguṇam |
saguṇaṃ kṛṣṇam akṛṣṇaṃ kṛṣṇo vyākhyāti sūryoktim ||

Manuscript:

18. Puṇḍarīka jyotiṣa 1. Ff. 1–27 and 29–53 (ff. 15–26 also numbered 1-12 in upper left margin of versos). 11 × 25 cm. 12/13 lines. Incomplete (adhyāyas 1–11; opening verses and final colophon omitted).

F. 1v begins, after salutations: atha sūryasiddhāṃtodāharaṇaṃ likhyate | tatra spaṣṭīkaraṇādau madhyamagrahasyopayogāt tasyāpy ahargaṇasādhyatvād ādāv ahargaṇaḥ sādhyate || saṃvat 1641 śālivāhanaśāke 1506 kārttikapaurṇamāsyāṃ śanau.

F. 27v ends in the ṭīkā on 4, 8: ābhyāṃ saumyo vikṣepaḥ 3|30 avaśi.

F. 29 begins in the ṭīkā on 4, 26(?): yasya katham api kathaṃ vā | kuto hetor ity artha ||

The text ends on f. 53: paścāt tāḥ ghaṭikā mānayogārddhaliptābhir guṇanīyāḥ pūrvānītakrāṃtyaṃtareṇa bhājyā labdhaghaṭikādi sthityarddhaṃ bhavati ||

After this is written:

daśāsyasya puryāḥ śarārkai⟨ḥ⟩ 125 kumārī
tato yojanaiḥ khāṣṭabhiḥ 80 kāṃtikāṃcī ||
tataḥ paṃcarāmai35ś ca mallāradhātrī
gajaiḥ 8 paryalī kheṃdubhi[ḥ]10r vatsagulmaṃ || 1 ||
khabāṇai[ḥ]50r avaṃtī kurukṣetram asyāḥ
kharudraiś ca 110 tattvāṣṭatulyaiḥ 825 sumeruḥ ||
bhaved aṃtaraṃ merulaṃkānagaryos
tathādhiṣṭhitaṃ jyotiṣādye bhacakraṃ || 2 ||
sarvayojanāni 1243 ||

F. 53v. Blank.

Another scribe has written upside down on f. 1r:

tithir na doṣo udaye ca kāle
nakṣatradoṣo abhijin na caivaṃ ||
yogo na doṣo ⟨'⟩stamaye ca kāle
na vāradoṣāḥ prabhavaṃti rātrau |

A third scribe has written in the margin of f. 29v:

śāko vedanavāśvi1294caṃdraviyuto ⟨'⟩dha⟨ḥ⟩stho tribhābhyāṃ 3/27 yuto sa [(bhayuktaḥ)] dvāviṃśatabhi⟨r⟩ 22 hata⟨ḥ⟩ śarādridhṛtibhi⟨r⟩ 1875 labdhāṣṭavedānvita⟨ḥ⟩ 48 ||
ṣaṣṭyā 60 tatpariśeṣitād iha samāśeṣaṃ kramāt tāḍayen māso dvādaśabhiḥ 12 khavahnibhi30r ahaḥṣaṣṭy⟨ā⟩ 60 ca daṃḍādayaḥ || 1 ||

The **GŪḌHĀRTHAPRAKĀŚAKA**, a ṭīkā on the *Sūryasiddhānta* composed by Raṅganātha at Kāśī in 1603. *CESS* A5, 388b–389a. Edited by Fitzedward Hall, Calcutta 1859. The first verse is:

yatsmṛtyābhīṣṭakāryasya nirvighnāṃ siddhim eṣyati ||
naras taṃ buddhidaṃ vande vakratuṇḍaṃ śivodbhavam ||

See manuscript 15. Another manuscript:

19. Khasmohor 5189. Ff. 1–197. 12 × 27 cm. 9/10 lines. Incomplete (adhyāyas 1–11).

Colophon on f. 197: iti śrīsakalagaṇakalagaṇakasārvabhauma vallāledevajñātmajaraṃganāthagaṇakaviracite gūṭārthaḥ kāśake pūrvakhaṃḍa paripūrtipag agamat.

After this is written: graṃthasaṃkhā catuḥsahasrā.

The **SAURABHĀṢYA**, a ṭīkā on the*Sūryasiddhānta* composed by Nṛsiṃha at Vārāṇasī in 1611. *CESS* A3, 204a–205a; A4, 162b–163a; and A5, 202b. The first verse is:

pratyudavyūhavidhvaṃsakāraṇāya mahātmane ||
gaṇeśāya namas tasmai jagatām eva sākṣiṇe ||

Manuscript:

20. Khasmohor 5188. Ff. 1–120. 12.5 × 27.5 cm. 10 lines. Normally only lemmata froṃ the mūla. Copied on *ca.* 4 July 1731.

Colophon on f. 120: iti śrīnṛsiṃhagaṇakaviracitaṃ sūryasiddhāṃtabhāṣyaṃ samāptam.

After this is written: graṃthasya saṃkhyā 'nuṣṭubhā puṣkaraviṣṇupadabhāvakṛṣṇavartmatulyā 310.

yādṛśaṃ pustakaṃ dṛṣṭvā tādṛśaṃ likhitaṃ mayā |
yadi śuddham aśuddhaṃ vā mama doṣo na dīyate ||

There follows the date-formula: saṃvat 1788 āṣyāḍhaśudī ekādaśī ke pustaka samāpta.

The **SAURAVĀSANĀ**, a ṭīkā on adhyāyas 1–11 of the*Sūryasiddhānta* composed by Kamalākara at Kāśī in *ca.* 1660. *CESS* A2, 23a; A4, 33b; and A5, 22a. Edited by Śrīcandra Pāṇḍeya, Vārāṇasī 1991. The first verse is:

> brahmāṇḍagolodaragaṃ khasaṃsthaṃ
> bhāntaṃ samastaṃ jalagolarūpam ||
> yattaijasaṃ bhāti yataḥ sadā taṃ
> nārāyaṇaṃ maṇḍalagaṃ namāmi ||

See manuscript 14. Another manuscript is:

21. Khasmohor 5592. Ff. 1–38 and 40–93. 11.5 × 26 cm. 8 lines. Copied by Nātha Bhaṭṭa, the son of Viṭṭhala Bhaṭṭa of the Medapāṭhajñāti, at Rūpapura in Gujarāta for the reading of the boys in the saṅgha of Mahārāja Jayasiṃha,on Saturday 31 August 1706. Incomplete.

F. 38v ends in the ṭīkā on 3, 22–24 (p. 43, 4): svachāyotthaka.

F. 40 begins in the ṭīkā on 3, 26c–27b (p. 43, 28): hnā(rkasya chāyā).

Colophon on f. 93: iti śrīsakalagaṇakasarvabhaumaśrīmannṛsiṃhātmajaśrīkamalākaraviracitā daśādhikārāṃtapūrvakhaṃḍasauravāsanā saṃpūrṇṇā.

Post-colophon: saṃvat 1763 varṣe bhādrapadamāse śuklapakṣe caturthi śanivāsarena gujarātamadhye rūpapuramadhye vāstavyaṃ medapāṭhajñātīyabha. vīṭhalasūtabha. nāthena laṣitaṃ pustakaṃ paṭhanārthaṃ māhārājādhirājagaubrāhmaṇapratipālaka chatrādhipatiśrīśrīśrīmahārājā jeyasaṃghaputrapaṭhanārthaṃ.

The **SIDDHĀNTAŚIROMAṆI** composed by Bhāskara at Vijjaḍaviḍa in 1150. *CESS* A4, 311b-319a, and A5, 258b-260a. The grahagaṇitādhyāya was edited by D. V. Āpaṭe with Bhāskara's own *Vāsanābhāṣya* and Gaṇeśa's *Śiromaṇiprakāśa* as *ASS* 110, 2 vols., Poona 1939-1941; the madhyamādhikāra with the *Marīci* of Munīśvara was edited by M. Jhā, Varanasi 1961; and the spaṣṭa to pāta with the *Marīci* of Munīśvara by K. Joshī, 2 vols, Varanasi 1964. The first verse is:

> yatra trātum idaṃ jagaj jalajinībandhau samabhyudgate
> dhvāntadhvaṃsavidhau vidhautavinaman niḥśeṣadoṣoccaye |
> vartante kratavaḥ śatakratumukhā dīvyanti devā divi
> drāṅ naḥ sūktimucaṃ vyanaktu sa giraṃ gīrvāṇavandyo raviḥ |

The golādhyāya was edited by D. V. Āpaṭe with Bhāskara's own *Vāsanābhāṣya* and Munīśvara's *Marīci* as *ASS* 122, 2 vols., Poona 1943-1952. The first verse is:

siddhiṃ sādhyam upaiti yatsmaraṇataḥ kṣipraṃ prasādāt tathā
yasyāś citrapadā svalaṅkṛtir alaṃ lālityalīlāvatī |
nṛtyantī mukharaṅgageva kṛtināṃ syād bhāratī bhāratī
taṃ tāṃ ca praṇipatya golam amalaṃ bālāvabodhaṃ bruve |

Manuscripts:

22. Khasmohor 5489. Ff. 1–17, 19–21 corrected to 18–20, and 21–84. 11 × 21.5 cm. 19 lines. With occasional marginalia. The grahagaṇitādhyāya with Bhāskara's *Vāsanābhāṣya*. Copied by Lakṣmaṇa Bhaṭṭa, the son of Kamalākara Bhaṭṭa, the son of Padmākara of the Jāmadagnigotra, on Thursday 22 January 1652. Formerly property of Lakṣmaṇa Josyā, who is probably the scribe.

Colophon on f. 84v: iti śrīmaheśvaropādhyāyasutaśrībhāskarācāryaviracite siddhāṃtaśiromaṇau vāsanābhāṣyamitākṣare pātādhyāyaḥ samāpto (yaṃ) graṃthaḥ.

After this is written: atrādhikāre graṃthasaṃkhyā catvāriṃśadadhikaṃ śatatrayaṃ 340 || samastagraṃthasaṃkhyā ṣaṭsārddhasahasrasaṃkhyā 6500.

There follows the date formula:

guṇaśailasāgarabhūmitaśāke śrīśālivāhanākhye ca ||
tapamāsi kṛṣṇapakṣe saptamyām iṃdrapūjite vāre || 1 ||
śrījāmagnyagotrodbhavaśrīmadpadmākarasya tanujo ⟨'⟩bhūt ||
śrīkamalākarabhaṭṭasutaḥ lakṣmaṇa[m] ālekhi vāsanābhāṣyaṃ || 2 ||

...ṣaṃvat 1708 śāke 1573 māghakṛṣṇa 7 gurau śrīmatpaṃḍitavaryapadmākarabhaṭṭasūnuśrīmatkama(lā)karabhaṭṭasūnunā
lakṣmaṇabhaṭṭena likhitā mitākṣarā.

Below this is written:

madhyamādhikāra	900
spaṣṭādhikāra	600
tripraśna	925
parvasaṃbhava	75
caṃdragrahaṇa	340
sūryagrahaṇa	325
chāyādhikāra	190
astodaya	100
śṛṃgonnati	310
grahayuti	85
bhagrahayuti	130
pātādhyāya	340
yogaḥ	4320
golādhyāy	2200
sarvayogaḥ	6520
iyaṃ sarvagraṃthasaṃkhyā	

23. Khasmohor 5488. Ff. 1–46. 11 × 21.5 cm. 16–21 lines. With some marginalia. The golādhyāya with Bhāskara's *Vāsanābhāṣya*. Copied by Lakṣmaṇa Bhaṭṭa, the son of Kamalākara Bhaṭṭa, the son of Padmākara Bhaṭṭa, on Sunday 21 September 1651.

Colophon on f. 46: iti śrīmaheśvaropādhyāyasutaśrībhāskarācāryaviracite siddhāṃtaśiromaṇo golādhyāyavāsanābhāṣye jyotpatyādhyāyaḥ.

Post-colophon: saṃvat 1708 śāke 1573 kharanāmasaṃvatsare kārttikakṛṣṇa 2 ravidine śrīmatpadmākarabhaṭṭasūnukamalākarabhaṭṭasūnunā lakṣmaṇena likhitam idaṃ pustakaṃ.

Below this is written a passate dated Wednesday 26 March 1617: atha kalpagataṃ likhyate ⟨|⟩ śālivāhanaśāke 1539 piṃgalābde pātaḥ ⟨|⟩ ṣaṇ manavaḥ 1840320000 ṣaṇmanūnāṃ sapta saṃdhyāsaṃdhyāṃśāḥ 12096000 ⟨|⟩ saptaviṃśati mahāyugāni 116640000 ⟨|⟩ kṛtayugaṃ 1728000 tretā 1296000 dvāpara 864000 ⟨|⟩ yudhiṣṭhiraśaka 3044 vikramaśaka 135 śālivāha⟨na⟩śake 1539 ⟨|⟩ trayāṇāṃ yoge gatakaliyugaṃ 4718 ⟨|⟩ evaṃ sarveṣāṃ yoge jātaṃ kalpagataṃ 1972948718 ⟨|⟩ asmād grahānayanaṃ yathā śake 1539 piṃgalābde caitrasuddha 1[1] budhe dhaniṣṭā 50 śobhana 27 di 31|44 madhyarātrau sūryasiddhāṃtād varttārakaḥ(?) ⟨|⟩ kalpagatam 1972948718 ⟨|⟩ tatra brahmaṇaḥ grahanakṣatrādisṛṣṭau gatakālāt divyābdāni 47400 ⟨|⟩ tāni ṣaṣṭyadhikaśatatrayagu360ṇitāni sauravarṣāṇi 17064000 ⟨|⟩ ete śodhyābdāḥ kal⟨p⟩agate śodhitā jātā sṛṣṭyādigatābdāḥ saurāḥ 1955884718 ⟨|⟩ asmāt ṣṛṣṭyādyahargaṇaḥ sādhyate.

On the bottom cover are written parameters of the *Sūryasiddhānta*.

grahabhagaṇāḥ	
ravibha	4320000000
caṃdrabha	57753336000
bhaumabha	2296832000
budhabha	17937060000
guru	364220000
śukra	7022376000
śani	146568000

uccabhagaṇāḥ		pātabhagaṇāḥ	
ra	387		
caṃ	488203000	caṃ	232238000
bhau	204	bhau	214
vu	368	vu	488
gu	900	gu	114
śu	535	śu	903
śa	39	śa	6620

nakṣatrodayāḥ	1582237828000
kudināni	1577917828000

caṃdrāhāḥ	1603000080000
adhimāsāḥ	1593336000
avamāḥ	25082252600
ravimāsāḥ	[1]51840000000
khakakṣāḥ	18712080864000000

ete yathā saṃbhavaṃ sahasrabhaktāḥ mahāyuge 4320000 bhagaṇāḥ syuḥ idaṃ bhagaṇādi brahmasiddhāṃtasūryasomaromaśavasiṣṭhavyāsasiddhāṃt⟨e⟩-bhyaḥ likhitaṃ

atha kalpapramāṇam 4320000000 tadyathā kṛta 1728000 tretāpra 1296000 dvāpara 864000 kalipra 432000 eṣāṃ caturṇāṃ yoge mahāyugaṃ 4320000 ebhir ekasaptabhi⟨r⟩ mahāyugair jāto manu[ḥ] pramāṇaṃ 306720000 caturdaśamanūnāṃ saṃdhyā saṃdhyāṃśāḥ kṛtapramāṇāḥ 1728000 etatsaṃdhyāsaṃdhyāṃśasahitacaturdaśamanūnāṃ yoge jātāni sahasramahāyugāni idam eva brahmadinaṃ kalpaś ca 4320000000.

24. Khasmohor 5593. Ff. 1–131 and 131b–134. 12 × 27.5 cm. 12 lines. Brittle paper; some leaves damaged. With some marginalia. The grahagaṇitādhyāya with Bhāskara's *Vāsanābhāṣya*. Copied for Satīdāsa, the son of Śrīdatta Jyotirvit, on Saturday 12 January 1657.

Colophon on f. 134: iti śrīmaheśvaropādhyāyasutabhāskarācāryaviracite siddhāṃtaśi⟨romaṇivāsanā⟩bhāṣye
mitākṣare pātādhyāyaḥ (the akṣaras within pointed brackets are lost in tears in this damaged leaf).

After this is written: gaṃthaḥ 440.

Post-colophon: jyotirvichrīdattātajasatīdāsapaṭhanā⟨rthaṃ saṃvat 1⟩713 śake 1578 pravarttamāne māghamāsīyaśuklapakṣe dvādaśyāṃ maṃdavāsare.

Price on top cover and in margin of f. 134v: kī 31 मु.

25. Khasmohor 5205. Ff. 1–25. 10 × 23.5 cm. 10/11 lines. With some marginalia. The yantrādhyāya from the golādhyāya with Bhāskara's *Vāsanābhāṣya*. Copied on Monday 3 June 1706. Formerly property of Jayakṛṣṇa Dīkṣita, the son of Śrīkṛṣṇa Dīkṣita.

F. 1 begins, after the salutation, with the heading to yantrādhyāya 1 (p. 360, 12 of vol. 2 of the Āpaṭe ed.): atha yaṃtrādhyāyo.

Colophon on f. 24v (*cf.* p. 427, 5–6): iti śrīmaheśvaropādhyāyasutabhāskarācāryaviracite siddhāṃtaśiromaṇī(corr. to °ṇi)vāsanābhāṣye mitākhye yaṃtrādhyāyaś caturdaśamaḥ.

After this is written: asmi[f. 25]nn adhyāye graṃthasaṃkhyā vi⟨ṃ⟩śatyadhikaṃ śatatrayaṃ 320.

Post-colophon: saṃvata 1763 śāke 1628 pravarttamāne āṣāḍhasudi caturthī 4 caṃdravāsare likhitaṃ.

Ownership note: dikṣata śrīkraṣṇasutadikṣataje kraṣṇasya pustakaṃ.

After this are enumerated members of this family: dikṣata jīvo dikṣata kaḍvo dikṣata kāśī dikṣata śrīvacaha dīkṣata jīvarāja dīkṣata śrīkraṣṇa dīkṣata jekraṣṇa.

Price on f. 25v: kī || – |||.

26. Khasmohor 5309. Ff. 1–45 and 47–181. 12 × 26.5 cm. 10–12 lines. Grahagaṇitādhyāya with Bhāskara's *Vāsanābhāṣya*. Copied by Tulārāma for Mahārājādhirāja Jayasiṃha on Thursday 8 September 1709.

Colophon on f. 181: iti śrīmaheśvaropādhyāyasutaśrībhāskarācāryaviracite siddhāṃtaśiromaṇau vāsanābhāṣya mitākṣare pātādhyāyaḥ samāptaḥ.

After this is written: atrādhikāre graṃthasaṃkhyā catvāriṃśadadhikaṃ śatatrayaṃ...

sarve te sukhinaḥ saṃtu sarve saṃtu nirāmayāḥ ⟨|⟩
sarve bhadrāṇi paśyaṃtu mā kaści⟨d⟩ duḥkham āpnuyāt ⟨||⟩ 1 ⟨||⟩

Post-colophon: śrīmanmahārājādhirājajīśrījayasighadevajīkasyājñayā likhitam idaṃ tulārāmeṇa ... saṃvat 1766 śake 1631 āśvinakṛṣṇa 1 gurau lekhaḥ.

27. Puṇḍarīka jyotiṣa 19. Ff. 1–5. 11 × 22.5 cm. 10 lines. Grahagaṇitādhyāya, madhyamādhikāra, kālamānādhyāya 15 with Munīśvara's *Marīci*. Copied by Dayārāma for Govinda Daivajña on Sunday 28 May 1727.

F. 1v begins, after the salutation: atha marīcibhāṣye kṛṣṇaśukrayor nirṇayam āha || There follows the heading to kālamānādhyāya 15 (p. 47, 3): atha kālasyānādyanaṃtatvād.

F. 5 ends (not in edition): pūrvācāryair varṣapravṛttir aṃgīkṛtā jātā dharmaśāstravacanād iti.

Colophon on f. 5v: iti marīcibhāṣye kṛṣṇādiśuklādinirṇayaḥ.

Post-colophon: saṃ 1784 āṣāḍhakṛṣṇa 5 ravau dayārāmeṇa lekhi śrīmaddaivajñagoviṃdasyārthe.

28. Khasmohor 5392. Ff. 1–131 (f. 52 is also numbered 23). 11 × 22 cm. 10 lines. With some marginalia. Golādhyāya with Bhāskara's *Vāsanābhāṣya*. Copied by Harinārāyaṇa Praśnorā, the son of Paṇḍita Joga, on Sunday 19 August 1733, probably at Kāśī.

Colophon on f. 131: iti śrīmaheśvaropādhyāyasutabhāskarācāryaviracite siddhāṃtaśiromaṇivāsanābhāṣye mitākṣare jyotpattiḥ samāptaḥ.

After this is written: śrīkāśiviśveśvarāpaṇam astu.

Post-colophon: saṃvat 1790 varṣe bhādirvā vadi 6 ravivāre liṣitaṃ paṃ jogātmajapraśnorā harinārāyaṇena.

29. Khasmohor 5030. Ff. 1–11. 11 × 20.5 cm. 8 lines. With some marginalia. The tripraśnādhikāra of the grahagaṇitādhyāya. Incomplete.

F. 1v begins, after the salutation, with the adhyāya heading (vol. 1, p. 134, 2 of Āpaṭe's edition): atha tripraśnādhyāyaṃ vivaktus tāvat tadārambhaprayojanam āha.

F. 11 ends, half through line 1, with 73d (vol. 1, p. 189, 17): cāpamo taḥ || 73 ||

30. Khasmohor 5391. Ff. 1–42. 11 × 22 cm. 8 lines. With marginal notes. Grahagaṇitādhyāya. Incomplete (ends in parvasambhavādhikāra).

F. 42v ends in parvasambhavādhikāra 2b (vol. 2, p. 2, 2 of Āpaṭe's edition): 169 yu.

31. Khasmohor 5439. Ff. 1–11 and 13–16 (f. 13 torn). 11 × 30 cm. 12 lines. Golādhyāya with Bhāskara's *Vāsanābhāṣya*. Incomplete (ends in the jyotpattivāsanā).

F. 11v ends in the heading to madhyagativāsanā (vol. 1, p. 115, 8 of Āpaṭe's edition): idānīṃ.

F. 13 begins in madhyagativāsanā 23a (numbered 22 in manuscript)(vol. 1, p. 126, 12): dayakālikās.

F. 16v ends in the ṭīkā on jyotpattivāsanā (vol. 1, p. 137, 23): trijyāyutena 100.

The **VĀSANĀBHĀṢYA** or **MITĀKṢARA**, a commentary on his own *Siddhāntaśiromoṇi* composed by Bhāskara at Vijjaḍaviḍa in *ca.* 1150. *CESS* A4, 319a–322a, and A5, 260a–261a. The first verse of the ṭīkā on the grahagaṇitādhyāya is:

> jayati jayati gūḍhānandhakāre padārthān
> janaghanaghṛṇayāyaṃ vyañjayann ātmabhābhiḥ |
> vimalitamanasāṃ sadvyāsanābhyāsayogair
> api ca paramatattvaṃ yogināṃ bhānur ekaḥ ||

The first verse of the ṭīkā on the golādhyāya is:

> golādhyāye nije yā yā apūrvā viṣamoktayaḥ |
> tās tā bālāvabodhāya saṃkṣepād vivṛṇomy aham ||

See manuscripts 22, 23, 24, 25, 26, 28, and 31. Other manuscripts:

32. Khasmohor 5245. A ff. 1–38, 40–46, 49–64, 63b, 63c, 64b, and 65–74 (text continuous); and B ff. 1–59, 63 (corrected to 59(b)), 60–62, 59c (corrected to 63), and 65 (ff. 16–32 also numbered 1–17; ff. 38–59 also numbered 1–22; f. 59b also numbered 24; and f. 65 also numbered 28). 11.5 × 28 cm. 7/8 lines.

Lemmata ususally from the mūla. Grahagaṇitādhyāya. Incomplete. B f. 65 apparently belongs to a different manuscript.

A f. 74 ends in the heading to tripraśnādhikāra 5 (vol. 1, p. 135, 7 of Āpaṭe's edition): idānīṃ lagnāt kālā.

B f. 1 begins in parvasambhavādhikāra 1a (vol. 2, p. 1, 7): kale gatābdā.

B f. 59c(corrected to 63)v ends in the ṭīkā on grahayutyadhikāra 7–9 (vol. 2, p. 119, 4): kiṃ tu yal la.

B f. 65 begins in the ṭīkā on madhyamādhikāra, pratyabdaśuddhi 10 (vol. 1, p. 52, 20): va śaśīty upapannaṃ.

B f. 65v ends, in the middle of line 7, in the ṭīkā on madhyamādhikāra, pratyabdaśuddhi 12a–b (vol. 1, p. 54, 5): śuddhitve.

33. Khasmohor 5390. Ff. 1–57, 57b–78, and 78b–142 (ff. 57v, 57b, and 78bv are blank). 10.5 × 22 cm. 10 lines. With marginal notes. Golādhyāya. Usually lemmata from the mūla.

Colophon on f. 142: iti śrīmaheśvaropādhyāyasutabhāskarācāryaviracite siddhāṃtaśiromaṇau vāsanābhāṣye mitākṣare golādhyāye jyotpatyadhyāyaḥ.

The **MARĪCI**, a commentary on Bhāskara's *Siddhāntaśiromaṇi* composed by Munīśvara Viśvarūpa at Kāśī in *ca.* 1638. *CESS* A4, 436b–438b, and A5, 314a. The first verse is:

> prātarbodhitapadmapatravikasannetro navīnāmbuda-
> śyāmaḥ pītadukūlaśobhitatanuḥ śrīvatsamudrānvitaḥ |
> vaṃśīvādanatatparo ravisutātīropago naicakī-
> gopībhiḥ pariveṣṭito mama manasyāstāṃ ramāvallabhaḥ ||

See manuscript 27.

The **SIDDHĀNTAŚIROMAṆYUDĀHARAṆA**, a commmentary on the grahagaṇitādhyāya of Bhāskara's *Siddhāntaśiromaṇi*, composed by Viśvambhara (*CESS* A5, 689a), known as Cakracūḍāmaṇi (*CESS* A3, 36b). The beginning is not known.

Manuscript:

34. Puṇḍarīika jyotiṣa 6. Ff. 4–42 and 44–96 (text continuous). 11 × 23 cm. 7 lines. Incomplete.

F. 4 begins in the ṭīkā on madhyamādhikāra, grahabhagaṇamānādhyāya 6a–b: caṃdrapātāḥ 2223111681 maṃ° pātāḥ 267.

Colophon on f. 96: iti śrīvājidaivajñacakracūḍāmaṇiviśvaṃbharaviracite siddhāṃtaśiromaṇyudāharaṇe pātādhikāraḥ.

The **BRAHMASIDDHĀNTA** alleged to be part of a *Śākalyasaṃhitā* and attributed to Brahman. *CESS* A4, 259a–260a, and A5, 240b–241a. Edited by V. Dvivedin as part 2 of fasc. 1 of his *Jyautiṣasiddhāntasaṅgraha*, *BSS* 39, Benares 1912. The first verse is:

dhyānayogasamārūḍhaṃ brahmāṇaṃ trijagadgurum |
abhivādya sukhāsīnaṃ nāradaḥ paripṛcchati ||

Manuscript:

35. Pothikhānā 25. Ff. A–B, 1–23, and C. 30.5 × 19.5 cm. 26/27 lines. European style book. With some marginal notes. Copied by Līchamīnārāyaṇa at Surata on *ca.* 24 August 1895. He was also the scribe of manuscripts 4, 5, and 17.

Ff. A–Bv. Blank.

Colophon on f. 23: iti śrīśākalyasaṃhitāyāṃ dvitīyaḥ praśne vrahmasiddhāṃte ṣaṣṭo dhyāya samāpta.

Post-colophon: ka° līchamīnārāṇa lekhata pothī° surata° mī° bhā° di 5 saṃ 1952.

Ff. 23v–Cv. Blank.

The **VṚDDHAVASIṢṬHASIDDHĀNTA = VIŚVAPRAKĀŚA** in the form of a dialogue between Vāmadeva and Vasiṣṭha. *CESS* A5, 608a–608b. The first verse is:

namas te viśvarūpāya parāya paramātmane |
yogidhyeyāya śāntāya kālarūpāya viṣṇave ||

Manuscript:

36. Puṇḍarīka jyotiṣa 4. Ff. 1–38 and A. 11 × 21.5 cm. 10 lines. Includes on ff. 35v–38v two chapters additional to the thirteen of the gaṇitaskandha: a golabandhādhyāya and an abhimānādhyāya. Copied on Friday 10 October 1690. Formerly property of Jagannātha Samrāt; purchesed by Viśveśvara, the son of Rāmeśvara ⟨of the Puṇḍarīka family⟩.

Colophon on f. 38v: iti śrībrahmārṣivṛddhavasiṣṭhapraṇīte || gaṇitaskandhaviśvaprakāśe bhimānādhyāyaḥ || || 15 ||

Post-colophon: saṃvat 1747 mitī kārttikakṛṣṇa 2 śukre taddine vasiṣṭhasiddhāṃtapustakaṃ samāptaṃ.

Ownership notes on f. 38v: idaṃ pustaṃ śrīsamrāṭjī jagaṃnnāthasya; on f. 1: idaṃ pustakaṃ śrīgajaṃnnātha sammrāṭjīkasya; and: pustakaṃjī || tatsakāśān mūlyena gṛhītaṃ

rāmeśvaratanūjasya viśveśvaravipaścitaḥ ||
vasiṣṭhasiddhāṃtapustakaṃ jagaj jānātu sarvadā || 1 ||

The **VYĀSASIDDHĀNTA** said to be the *Vedāṅgajyotiḥśāstra* in a *Vyāsasmṛti*. *CESS* A5, 754a–754b. Edited by Radhavallabh Jyotiṣatīrtha, Calcutta [N. D.]. The first verse is:

bhavaṃ nārāyaṇaṃ sākṣāt praṇipatya kṛtāñjaliḥ |
bhaktyā paramayābhyarcya papracchedam ṛṣis tadā ||

Manuscript:

37. Khasmohor 5257. Ff. 1–12. 11.5 × 27 cm. 12 lines.

Colophon on f. 12v: iti śrīvyāsakṛtau dharmaśāstre vyāsasmṛtau vedāṃgajyotiḥśāstre golapratipādanaṃ samāptaṃ || || śloka 275 iti vyāsasiddhāṃtaḥ samāptaḥ.

The **SOMASIDDHĀNTA** in the form of a dialogue between Śaunaka and Soma. *CESS* A6. Edited by V. Dvivedin as part 1 of fascicle 1 of his *Jyautiṣasiddhāntasaṅgraha*, Benares 1912. The first verse is:

bṛhaspatisutaṃ śāntaṃ sukhāsīnaṃ priyekṣaṇam |
abhivandyaṃ munir dhīmān śaunakaḥ paripṛcchati ||

Manuscripts:

38. Museum 47. Ff. 1–19. 8 × 28.5 cm. 6 lines. Copied on *ca.* 6 March 1679.

Colophon on f. 19v: iti śrīsomasiddhāṃte caturtho golādhyāyo daśamaḥ || 10 || samāptim agamad ayaṃ graṃthaḥ || || samāpta.

Post-colophon: saṃvat 1736 śamae caita sudi 5 paṃcami.

39. Puṇḍarīka jyotiṣa 3. Ff. 1–18. 11 × 23.5 cm. 10 lines. Copied on Sunday 13 October 1723. Formerly property of Jagannātha Samrāṭ; purchased by Viśveśvara, the son of Rāmeśvara ⟨of the Puṇḍarīka family⟩.

Colophon on f. 18: iti śrīsomasiddhāṃte caturtho golādhyāyo daśamaḥ || || samāptim agamad ayaṃ graṃthaḥ.

Post-colophon: saṃvat 1780 mitī kārttikakṛṣṇa 10 ravivāre taddine samāptaḥ.

Ownership notes on f. 18: idaṃ pustakaṃ śrījagannāthasamrāṭjīkasya; and on f. 1: idaṃ pustakaṃ śrīsamrāṭjagannāthasya; and:

rāmeśvaratanūjasya viśveśvaravipaścitaḥ ||
jagaj jānātv idaṃ pustaṃ || || golādhyāyasya sarvada || 1 ||;

then, below śrīsamrāṭjagannāthasya: tatsakāśāt maulyena gṛhītaṃ.
f. 18v. Blank.

40. Khasmohor 1620 Ff. 1–17. 13.5 × 34.5 cm. 9 lines. With some marginalia; some corrections in pencil.
Colophon on f. 17: iti śrīsomasiddhāṃte caturtho golādhyāyo daśamaḥ 10.

The **SUNDARASIDDHĀNTA** composed by Jñānarāja at Pārthapura in 1503. *CESS* A3, 75a–76b; A4, 100a; and A5, 122b–123a. The first verse of the grahagaṇitādhyāya is:

diṅmātaṅgasutuṅgapañcavadanaṃ viśvaikalambodaraṃ
cūḍāratnasahasrabhūdharamahāhāraṃ sunīlāmbaram |
svāntadhvāntaharaṃ kalānidhidharaṃ koṭīniruksundaraṃ
vārāhopamavāhanaṃ gaṇapatiṃ vande paraṃ śaṅkaram ||

The first verse of the golādhyāya is:

bhāle yasya kalānidhir madhumiladbhṛṅgāvalīgaṇḍayoḥ
kaṇṭhe 'hir vilasatpalaṃ padayuge gīrvāṇacetoguṇaḥ |
brahmāpi trijagat sisṛkṣur alaṃ nirvighnasaṃsiddhaye
śrīmanmaṅgalamūrttim ādyam alayaṃ taṃ naumi bhaktapriyam ||

Manuscript:

41. Khamsmohor 5591. Ff. 1–63. 12 × 26 cm. 8 lines. Golādhyāya (incomplete) on ff. 1v–32, grahagaṇitādhyāya on ff. 32–63. Copied by Tulārāma for Mahārājādhirāja Jayasiṃha on Tuesday 5 November 1706.
F. 11 ends, in the middle of line 1, in golādhyāya, bhuvanakośa 77c: sāmujyatāṃ bhagavato.
The rest of f. 11 and ff. 11v–12. Blank.
F. 12v begins in golādhyāya, madhyabhuktivāsanā 9d: yojyaṃ tadbhramaṇaṃ.
Colophon on f. 32: eṣa golādhyāyaḥ.
Final verse on f. 63:

itthaṃ śrīmannāganāthātmajena
prokte tantre jñānarājena ramye ||
granthāgārādhārabhūte prabhūte
śṛṅgonnatyadhyāya evaṃ niruktaḥ ||

"Post-colophon" (there is no final colophon): saṃ 1763 varṣe kārttikaśukla 11 bhaume || || śrīmanmahārājādhirājaśrījayasiṃhadevajīkasyājñayā (suṃdarasiddhāṃtanāmā graṃtho) lekhi | tulārāmeṇa.
After this is written:

akṣaramātrapadasvarahīnaṃ
vyaṃjanasaṃdhivivarjitarephaṃ ||
sādhubhir eva mama kṣamitavyaṃ
ko ⟨'⟩tra na muhyati śāstrasamudre || 1 ||

F. 63fv. Blank.
Price on f. 1: kī || ≡.

The **SIDDHĀNTATATTVAVIVEKA** composed by Kamalākara at Kāśī in 1658. *CESS* A2, 21a–22b; A3, 18a; A4, 33a–33b; and A5, 22a. Edited by S. Dvivedin and M. Jhā, reprinted Vārāṇasī 1991. The first verse is:

brahmāṇḍodaramadhyagāvanijalāgnyūrdhvendupūrvagraha-
rkṣordhvasthapravahāntagolaracanā sṛṣṭir yathāvat sthitā |
kāle 'smin gahane 'vyaye 'sti satataṃ yasmād iyaṃ taj jayaty
ādyaṃ nirguṇam īśam avyayaparabrahmaikatattvaṃ śubham ||

Manuscripts:

42. Khasmohor 5301. Ff. 1–26. 11.5 × 26.5 cm. 10 lines. Incomplete (ends with madhyamādhikāra).
Colophon on f. 26v: iti śrīkamalākaraviracite siddhāṃtatattvaviveke madhyamādhikāraḥ || || 1 ||
Price on f. 1: kī ||ȷ.

43. Khasmohor 5303. Ff. 1–72. 12 × 29 cm. 8 lines. Spaces left blank for diagrams that were never drawn. Incomplete.
F. 72v ends in a prose passage that seems to be connected to the long section of the spaṣṭādhikāra devoted to Sines: atha paṃcamāṃśājñena ca paṃcaguṇitabhujāṃśānāṃ jyāsvarūpa...bhujajyāvargonas trijyāvargaḥ koṭijyāvarga iti (*cf.* spaṣṭādhikāra 56 on. p. 108) ...atra paṃcamāṃśarūpakevalabhujajyājñānārtham idaṃ jñātajyāsamam iti...paṃcaguṇitena trijyāvargavarge (here it ends).
Price on f. 1: kīmatī|| ≡ ||S.

The **SŪRYASIDDHĀNTASĀRAVICĀRA**, a treatise on planetary models and spherical trigonometry based on the *Sūryasiddhānta*, composed, it is alleged, by Samrāṭ Jayasiṃha, presumably at Amber in *ca.* 1715–1720. This is attributed hesitatingly to Jagannātha Samrāṭ in *CESS* A5, 114a. The Jayapura manuscript begins, after the salutation: atha spaṣṭādhikāre tripraśnādau ca sūryasiddhāṃte kiṃcid yathāmati vicāryate. In the upper margin of f. 1v are written by a second scribe two verses:

ānaṃdavallikaṃdaṃ kaṃdadyutivṛṃdamaṃdiram amaṃdam ||
raghunaṃdanam aṃtar ahaṃ vaṃde vaṃdārumaṃdāram || (|| 1 ||)
śrīmadbhiḥ samrāḍbhir jayasiṃhanṛpāgraṇībhir atigūḍhaṃ ||
spaṣṭādhikṛtau spaṣṭaṃ vivecyate golavijñajanatuṣṭyai || 2 ||

These are the first two verses of the *Sūryasiddhāntasāravicāra* in its only other identified copy, RORI 29498. On f. 1 of the Jaipur manuscript the same scribe who copied these two verses of f. 1v has written: sūryasiddhāṃtasāravicāraḥ śrīmahārājakṛtaḥ. Moreover, the words śrīmadbhiḥ samrāḍbhir jayasiṃhanṛpāgraṇībhir ullikhitā occur within the text (f. 22 of the Jaipur manuscript).

Manuscript:

44. Khasmohor 4955. Ff. A–B, 1–8, C–D, 9–19, 19b, 20–26, 27/28, 29–56, 56b, 57–73, E–F, 74–79, and G. 12 × 29 cm. 7–9 lines. With marginal notes. A number of diagrams have been added to the manuscript, but their relation to the text remains obscure. They were drawn and lettered by the scribe who added the two verses on f. 1v and who was responsible for the marginal notes.

F. A. Blank.

Ff. Av–B. Diagram of a planetary model in which the manda epicycle produces an eccentric circle to act as the deferent of the center of the śīghra epicycle.

F. Bv. Blank.

F. C. Blank.

Ff. Cv–D show a model in which the center of the manda epicycle travels on a deferent whose center is carried on the circumference of a small circle whose center is the earth; this is similar to the Ptolemaic crank-mechanism of the Moon save that the center of the deferent and the center of the manda epicycle rotate in the same direction. The center of the śīghra epicycle rotates on the circumference of the manda epicycle.

F. Dv. Blank.

F. 19b has a diagram for the Sun with a manda epicycle whose center rotates on a concentric deferent while the center of a secondary (labeled śīghra) epicycle rotates on the circumference of the manda epicycle.

F. 19bv. Blank.

F. 41v has an unclear diagram referring to the mandakarṇas of the Sun and the Moon.

F. 45v has an unclear diagram relating to the mandakarṇas of the superior planets.

F. 51v has two diagrams: one relates to deflection (valana), and the other to the mandakarṇas of the Sun and the Moon.

F. 56b has a diagram illustrating retrogression according to a geocentric model.

F. 56bv. Blank.

F. 64v. Blank.

F. 73v has two diagrams illustrating an eccentric circle.

F. E has two diagrams relating to planetary and lunar latitude respectively.

F. Ev has a diagram showing the two epicycles of which the larger has no deferent.

F. F has a diagram showing the equivalence of a manda epicycle and an eccentric deferent.

F. Fv has a diagram of a model for the two inferior planets in which the center of the manda epicycle rotates on the circumference of a concentric deferent and the center of the śīghra epicycle on the circumference of the manda epicycle.

F. 79v ends, a third of the way through line 7: yad vā svalpāṃtaratvād rad raviṇā sūkṣmāṃnād ata iti dik.

There is no colophon.

F. G. Blank.

Price on f. 1 and f. Gv: kī 3ȷ.

The earliest version of the **SIDDHĀNTAKAUSTUBHA** composed by Jagannātha Samrāṭ at Amber in about 1727. Some of the manuscrits listed in *CESS* A3, 57a–58a, and A5, 114a, undoubtedly contain the *Siddhāntakaustubha* instead of or as well as the *Samrāṭsiddhānta* that is a translation of Ptolemy; among these are RORI 17213, 19747, and 36323. A second version of this text was edited by M. Caturveda, Sāgara 1976 (it mantions 1730); and a third by R. Śarman, *Samrāṭsiddhānta*, vol. 2, Naī Dillī 1967, pp. 1029–1213 (it mentions 1732). The opening verses of the earliest version are:

gaṇādhipaṃ surārcitaṃ samastakāmadaṃ nṛṇāṃ |
praśastabhūtibhūṣitaṃ smarāmi vighnavāraṇam || 1 ||
lakṣmīnṛsiṃhacaraṇāmburuhaṃ sureśair
vandyaṃ samastajanasevitareṇugandham |
vāgdevatāṃ nikhilamohatamopahantrīṃ
vande guruṃ gaṇitaśāstraviśāradaṃ ca || 2 ||
śrīgovindasamāhvayādivibudhān vṛndāṭavīnirgatān
yas tatraiva nirākulaṃ śucimanobhāvaḥ svabhaktyānayat |
mlecchān mānasamunnatān svatarasā nirjitya bhūmaṇḍale
jīyāc chrījayasiṃhadevanṛpatiḥ śrīrājarājeśvaraḥ || 3 ||
karaṃ janārdanaṃ nāma dūrīkṛtya svatejasā |
bhrājate duḥsaho 'rīṇāṃ yathā graiṣmo divākaraḥ || 4 ||
yeneṣṭaṃ vājapeyādyair mahādānāni ṣoḍaśa |
dattāni dvijavaryebhyo gogrāmagajavājinaḥ || 5 ||
tena śrījayasiṃhena prārthitaḥ śāstrasaṃvidā |
karoti śrījagannāthaḥ samrāṭ siddhāntakaustubham || 6 ||
kṛtī jayati bhāskaro gaṇitagolatattvārthavin
munīśvara udāradhīḥ prathitatantrasadvāsanaḥ |
jayanti kamalākarādaya udārasadyuktayaḥ
kṛtī bhavati mādṛśaḥ samavalokya yeṣāṃ kṛtīḥ || 7 ||

Manuscript:

45. Puṇḍarīka jyotiṣa 5. Ff. 1–38. 11 × 22 cm. 9–12 lines. With some marginalia. Incomplete.

F. 38v ends, in the middle of line 8, with verse 60 of a section entitled akṣakṣetrāṇi:

> chāyādvayaṃ caikakapāla eva
> tadā bhavet kālavibhedasiddham |
> aśodhitākṣāṃśavaśāt tathātra
> prabhā bhaved yāmyadig aṃśakeṣu || 60 ||

C. KARAṆAS

The karaṇa version of the **RĀJAMṚGĀṄKA** composed by Bhojadeva (Bhojarāja) at Dhārā in 1042. *CESS* A4, 337a, and A5, 266b, where the koṣṭhaka version is included. Edited by D. Pingree in *Aligarh Oriental Series* 7, Aligarh 1987, pp. 4–42.

The first verse is:

arkaś candraḥ kujaḥ saumyo jīvaḥ śukraḥ śanis tamaḥ |
ketur grahā navāpy ete pāntu vo duritacchidaḥ ||

Manuscript:

46. Khasmohor 5589. Ff. 1–22. 12 × 27 cm. 10–12 lines. With some marginal notes. Copied for his own use by Śrīdatta Jyotirvit, the son of Śaṅkara, the son of Puruṣottama Vyāsa, a resident of Jodhapura, on Thursday 25 June 1618. The text differs considerbly from the edition.

Edition	Khasmohor 5589
madhyama 22 verses	madhyamānayana 57 verses
spaṣṭīkaraṇa 68 verses	sphuṭagati 223 verses
tripraśna 57 verses	tripraśna 43 verses
udayāsta 100 verses	candragrahaṇa 18 1/2 verses
śṛṅgonnati 11 verses	sūryagrahaṇa 11 verses
candragrahaṇa 65 verses	udayāsta 14 verses
sūryagrahaṇa 56 verses	grahayuti 24 verses
grahayuti 91 verses	śṛṅgonnati 27 verses

Colophon on f. 22v: iti śrīmahārājādhirājaparmeśvarasakalavilāsinījīviteśvaraśrībhojadevaviracite
śrīrājamṛgāṃkanāmni karaṇe śṛṅgonnatyadhikāro 'ṣṭamaḥ saṃpūrṇaḥ || samāptaṃ rājamṛgāṃkakaraṇam iti.

Post-colophon: saṃvat 1675 varṣe śāke 1540 āṣāḍhasudi 13 gurau vyā° puruṣottamasutasaṃkaratatputrajyotirvit śrīdattena likhitam idaṃ ātmapaṭhanārthaṃ || (|| jodhapuranayare vāstavyaṃ ||).

Price on f. 1: kī ≡.

The **KARAṆAPRAKĀŚA** composed by Brahmadeva at Mathurā in 1092. *CESS* A4, 257b–258b, and A5, 240a–240b. Edited by S. Dvivedin as *CSS* 23, Benares 1899. The first verse is:

brahmācyutatrinayanārkaśaśāṅkabhauma-
saumyejyaśukraśanivāgadhipāgaṇeśān |

natvāham āryabhaṭaśāstrasamaṃ karomi
śrībrahmadevagaṇakaḥ karaṇaprakāśam ||

Manuscripts:

47. Khasmohor 4953. Ff. 1–13. 11 × 24 cm. 9/10 lines. Copied by the son of Keśava on a Friday in 1626/7.

Colophon on f. 13v: iti śrīvrahmadevaviracite karaṇaprakāśe grahayutyadhikāraḥ samāptaḥ.

Post-colophon:

aṣṭasāgaraśilīmukhaśrubhiḥ 1548
saṃmite śakavare bhṛguvāre
keśavasya tanayaḥ paripūrṇaṃ
vrahmadevakaraṇaṃ samakārṣīt

After this is written:

ālolatulaśīmālam ārūḍhavinatīsutaṃ ||
jyotiriṃdīvaraśyāmam āvir astu mamāgrataḥ ||

48. Puṇḍarīka jyotiṣa 33. Ff. 1–12. 11 × 24 cm. 9/10 lines. A careful copy of manuscript 47, including its colophon, post-colophon, and additional verse. Formerly property of Viśveśvara, the son of Rāmeśvara Puṇḍarīka.

Colophon on f. 12v: iti śrīvrahmadevabiracite karaṇaprakāśe grahayuddhadhikāraḥ samāptaḥ.

Post-colophon:

aṣṭasāgaraśilīmukhaśrubhiḥ 1548
saṃmite śakavare bhṛguvāre
keśavasya tanayaḥ paripūrṇaṃ
vrahmadevakaraṇaṃ samakārṣīt

After this is written:

ālolatulaśīmālam ārūḍhavinatāsutaṃ ||
jyotiriṃdīvaraśyāmam āvir astu mamāgrataḥ ||

Ownership verse on f. 1:

rāmeśvaratanūjasya viśveśvaravipaścitaḥ ||
karaṇaprakāśapustaṃ jagaj jānātu sarvadā || 1 ||

The **BHĀSVATĪ** composed by Śatānanda at Puruṣottama in 1099. *CESS* A6. Edited by R. Miśra as *KSS* 46, 2nd ed., Vārāṇasī 1985. The first verse is:

natvā murāreś caraṇāravindaṃ
śrīmāñ śatānanda iti prasiddhaḥ |
tāṃ bhāsvatīṃ śiṣyahitārtham āha
śāke vihīne śaśipakṣakhaikaiḥ ||

Manuscripts:

49. Khasmohor 5408. Ff. 1–8. 11.5 × 23 cm. 10 lines. Ff. 1–2 and 7–8 are double folios. Folios much damaged. With extensive marginalia. Copied in 1606/7.
Colophon on f. 8: iti śrīpaṃcasiddhāntasame bhāsvatīkaraṇe parilekhādhikāro ṣṭamaḥ
Post-colophon: saṃvat 1663.
Price on f. 1: kī = |S.

50. Khasmohor 5358. Ff. 1–10. 11 × 25 cm. 17 lines. With an udāharaṇa. Incomplete (adhikāra 8 omitted). Copied on *ca.* 23 October 1699.
Colophon on f. 10v: iti śatānaṃdaviracite bhāsvatīkaraṇe sūryagrahaṇādhikāraḥ | samāptitaṃ bhāsvatīkaraṇaṃ sodāharaṇam idaṃ.
Post-colophon: saṃvat 1756 kārttikasudi 10.

51. Khasmohor 5572. Ff. 1–2 and 4–9. 11 × 12 cm. 13/14 lines. With marginalia. Incomplete. Copied for himself by Dhanarāja, the son of Vidyāvinoda Mathena, on Saturday 17 May 1701.
F. 2v ends in 2, 3a (p. 13, 2): abdaḥ pṛthak kheśa110guṇaḥ.
F. 4 begins in 3, 2d (numbered 3 in manuscript)(p. 33, 8): rgaṇa māṃśakaḥ saṃyutaḥ syāt 3.
Colophon on f. 9v: iti śrībhāsvatyāṃ parileṣādhikārāṣṭamaḥ samāpta || iti śrīsaurapakṣe bhāsvatīgraṃtha samāpta.
Post-colophon: saṃvat 1758 varṣa jyeṣṭavadi 6 ravisute mathenavidyāvinoda tatputra dhanarājene likhitaṃ svavācanārthaṃ.
After this is written: graṃthāgraṃtha śloka 115.

52. Khasmohor 5195. Ff. 1–12. 11.5 × 26.5 cm. 8 lines. Copied by Tulārāma at Amadāvāda (Ahmadābād) for Mahārājādhirāja Jayasiṃha on Sunday 8 December 1706.
Colophon on f. 12: iti bhāsvatīkaraṇe parilekhādhyāyaḥ samāptaḥ.
Post-colophon on f. 12v: saṃ 1763 varṣe mārgaśīrṣaśukla 15 ravau śrīmanmahārājādhirājajīśrījayasiṃhadevajīkasyājñayā likhitam idaṃ tulārāmeṇa

|| || mukāma amadāvādamadhye.
Price on f. 1: kī |J.

53. Museum 176. Ff. 1–8. 9.5 × 25 cm. 10 lines. With many marginalia. Copied by Rāmadatta Rāma in 1791/2.
Colophon on f. 8: iti śrīśatānaṃdaviracite bhāsvatīkaraṇe parilekhādhikāraḥ.
Post-colophon: samva 1848 rāmadatarāmena līkhīte.
This is followed by:

karko 'gne yama nairṭe ca vāyudiśā jñāyate kathitaṃ
īśānottaradik tathā ca kāryaṃ taṃ tūryye diśā tūryabhuk
ity etat kathitaṃ varāhamihire rudre tathā yāmale
saṃdehasya vināśanāya bhavitā rudroktapriyā cāgvataḥ(?) || 1 ||

54. Khasmohor 5305. Ff. 1–35. 10/11 × 24.5 cm. 7 lines. With corrections and marginal notes by a second scribe. With an udāharaṇa. Incomplete (adhikāra 8 omitted).
Colophon on f. 35: iti bhāsvatīkaraṇe sūryagrahaṇādikāra samāpto yam iti.
Price on f. 1: kī | = |||S.

A **BHĀSVATYUDĀHARAṆA**, an anonymous commentary on Śatānanda's *Bhāsvatī*. It begins: tatrādau sakalavighnopaśamanārtham iṣṭadevatānamaskārapūrvaṃ sambandhādhikam āha. It continues after verse 1: asyānvayaḥ || śrīmān śatānanda iti prasiddhaḥ tāṃ bhāsvatīm āha.

See manuscripts 50 and 54.

The **KARAṆAKUTŪHALA** composed by Bhāskara at Vijjaḍaviḍa in 1183. *CESS* A4, 322a–326a, and A5, 261b–263a. Edited with the ṭīkā of Sumatiharṣa by S. Miśra, Vārāṇasī 1991. The first verse is:

gaṇeśaṃ giraṃ padmajanmācyuteśān
grahān bhāskaro bhāskarādīṃś ca natvā |
laghuprakriyaṃ prasphuṭaṃ kheṭakarma
pravakṣyāmy ahaṃ brahmasiddhāntatulyam ||

Manuscripts:

55. Khasmohor 5057. Ff. 1–8. 11.5 × 27 cm. 11–13 lines. With marginalia. Copied by Pāṇḍai(?) on *ca.* 10 May 1572.
F. 8 ends with 11, 6d (p. 121, 4): kuśamaṃyuli arghayauḥ.
Colophon on f. 8: iti abhrānayana samāptaḥ.
Post-colophon: saṃ 1629 jeṭhavadi 13 li° pāṃḍai.

Below this is written the table of bhogas 1 to 4 of the Sun from Bhojadeva's *Rājamṛgāṅka*. The entries are for the epoch (10^s 28; 45, 0°); 1, 2, 3...9 days; 10, 20, 30...90 days; 100, 200, 300...900 days; and 1000, 2000, 3000...9000 days. The heading is: rājamṛgāṃke ravibhogā likhyaṃte.

56. Khasmohor 5203. Ff. 1–25. 11.5 × 20 cm. 8 lines. Incomplete (adhyāya 11 missing). Copied by Līlādhara near the Mahākāleśvara temple in Avantikā (Ujjayinī) on Tuesday 18 August 1705 during the reign of Mahārāja Jayasiṃha.

Colophon on f. 24: iti śrīpaṃḍitabhāskaraviracite brahmasiddhāṃtatulye karṇakautūhalẹ daśamadhikāra grahaṇasaṃbhavaṃ samāptaṃ || śrī || samāpta graṃtha.

After this is written:

gataiṣyanāḍīguṇitā[a] khakhāṣṭaiḥ[b] 800
sarvarkṣanāḍīvihṛtā[c] kalādyā[ḥ] |
bhuktarkṣayu⟨ktāḥ⟩[d] sakalā grahāś ca
ṣaṣṭyāhate ṣaṣṭiśateṣu(?) bhukti⟨ḥ⟩ || 1 ||

a. °nāḍiguṇitā. b. khakhāṣṭau. c. sarvakṣanāḍivihṛtā. d. bhuktirkṣayu.

After this is another colophon: brahmatulyamūla saṃpūrṇaṃ.

Post-colophon: saṃvat 1762 varṣe bhādrapadaśuklapakṣe 10 maṃdavāsare li° līlādhareṇa || śubham astu | avaṃtikāmadhye māhākāleśvarakṣiprāsannidhau idaṃ pustakaṃ ||

yādṛśaṃ pustakaṃ dṛṣṭvā tādṛśaṃ likhitaṃ mayā ||
yadi śuddham aśuddhaṃ vā mama dośo na dīyate || 2 ||
tailād rakṣe jalād rakṣed rakṣe śrathilabaṃdhanāt |
mūrṣahaste na dātavyaṃ evaṃ vadati śāradā || 2 ||

mahārājaśrī | jayasiṃghajīvijayarājye.

57. Khasmohor 5356. Ff. 1–7. 11.5 × 26 cm. 14 lines. Incomplete (adhyāya 11 omitted). Copied on *ca.* 30 March 1711.

F. 7v ends with 10, 10a–c (p. 119, 6–7):

itīha bhāskarodite grahāgame kutūhale
vidagdhabuddhivallabhe

Colophon on f. 7v: iti śrībrahmatulye pātasaṃbhavādhikāro daśama | 10 ||.
Post-colophon: saṃvat 1768 varṣe vaiśākhavadi 8 dine likhitaṃ.

58. Khasmohor 5583. Ff. 1–20. 11.5 × 27 cm. 7 lines. With extensive marginalia by a second scribe. Incomplete (adhyāya 11 omitted). Copied by

Tejabhāna on Tuesday 10 February 1756. Formerly propety of Tejabhāna, the son of Gulābrāya.

After the colophon to adhyāya 10 is written on f. 20:

> gaṇodhas(?) triyuk svākṣigogāṃgayukta692s
> triṣaṭ63bhakta āptāvamair yukta ūrddhaḥ
> kharāmai30r hataḥ saikaśeṣaṃ tithiḥ syāt
> phalaṃ māsavṛd(?) aṃtato dho dvinighnāt 1
> rasāṃgānvitāt sve bhanetrāṃka928labdhā
> vihīnā dagāṃgā(?) 66 bhānona(?) ūrddhaḥ
> hato bhānubhiḥ 12 śeṣakaṃ yātamāsā
> gatābdā phalaṃ śeṣukheṣaṃ 1105 śakas tat 2

Colophon: iti śrībhāskarācāryakṛte brahmatulyakaraṇam.

Post-colophon:

> dvyekaṣṭeṃdumite varṣe māghe ṃtakatithau sudi
> kuje mayedaṃ likhitaṃ tejabhānena pustakaṃ |

saṃbat 1812 śakaḥ 1677 māghasudi 10 bhaume.

Ownership note on f. 1:

> śrīmadgulābrāyasya tejabhāno stai puttrakaḥ
> tasyedaṃ pustakaṃ jñeyaṃ khanetramitapattrakam.

59. Khasmohor 5427. Ff. 1–13. 11 × 26 cm. 9/10 lines. Copied on Monday 21 August 1769.

Colophon on f. 13v: iti śrībrahmasiddhāṃtatulye bhāskarīye karṇakutūhale parvānayanādhikāro daśamaḥ 10.

Post-colophon: saṃvat 1826 bhādre kṛṣṇe 4 caṃdre li°.

After this is written adhyāya 11, ending with 11, 6d (p. 121, 4): kusumāṃjalinārcayet.

60. Puṇḍarīka jyotiṣa 40. Ff. A and 1–10. 9.5 × 24.5 cm. 11 lines. Incomplete (adhyāya 11 omitted). Copied by Gopīnātha, the son of Rāmeśvara Vyāsa, on Friday 31 October 1788. Formerly property of Viśveśvara, the son of Rāmeśvara Puṇḍarīka.

Ff. Av–1r. Blank.

Colophon on f. 10v, after 10, 4 (p. 119, 5): iti karṇakutūhalaṃ samāptaṃ.

Post-colophon: likhitaṃ vyāsopanāmakarāmeśvarasyātmajagopīnāthena || || śubhaṃ bhavatu || || saṃvat 1845 kārttika śukra 2 bhṛgau.

Ownership note on f. A:

> rāmeśvaratanūjasya viśveśvaravipaścitaḥ ||
> jñeyaṃ niraṃtaraṃ sadbhir brahmatulyasya pustakaṃ || 1 ||

61. Khasmohor 5144. Ff. 1–30, 29 (corrected to 31), 31b, 31c (renumbered 32), 32b–45, 45b (corrected to 46), and 47–64. 13 × 24 cm. 10 lines. With the ṭīkā of Padmanābha. Incomplete.

F. 64v ends, in the middle of line 8, in the ṭīkā on 5, 8 (p. 69, 20–21): evaṃ yadi sakṛt kṛte laṃbanave.

Price on f. 1: kīmatī ¶||–.

62. Khasmohor 5175(b). 1 f. 11.5 × 28.5 cm. 9 lines. Incomplete (excerpts).

F. 1r. paṃcayugābdhi445hīne śake ṣaṣṭyā 60 hṛte ayanāṃśāḥ | 1 | karṇa-kautūhale || There follow *Karaṇakutūhala* 1, 2–3 (p. 2, 16. śaka to 19. yaṃ ||); 1, 4d (p. 4, 24. navā9tyaṣṭi to caṃdrapāte ||); and 1, 9c–d (p. 9, 13. dvidhāṃkacaṃdrai to bhavatīṃdupātaḥ ||).

F. 1v begins in 3, 1a (p. 35, 3): laṃkodayā.

F. 1v ends in 3, 7d (p. 38, 18): conataṃ yā.

63. Khasmohor 5406. Ff. 1–43 (f. 8r blank; no text lost). Copied by two scribes: α ff. 1v–19v and β ff. 20r–43v. Dimensions not recorded. α 8–12 lines, and β 10 lines. With an anonymous vivṛti. Incomplete.

α f. 19v ends in the ṭīkā on 3, 2c–4d (p. 35, 6–10): lagnabhuktakālāna(yana)..

β f. 20 begins in the ṭīkā on 3, 5 (p. 36, 29–30): yor yogaḥ 300 madhyo-dayābhāvād ayam eveṣṭakālo lagnāj jātaḥ.

β f. 43v ends in the ṭīkā on 9, 1a–2b (p. 103, 3–5): samāpamatve ravicaṃdrayoḥ krāṃtisāmye sati tadaikye.

Price on f. 1: kī ||| · |||S.

The **NĀRMADĪ**, a commentary on Bhāskara's *Karaṇakutūhala* composed by Padmanābha in *ca.* 1400. *CESS* A4, 170a–170b, and A5, 205b. The first verse is:

jñātum saduktilavayugvivṛtiṃ tu śiṣyaiḥ
saṃprārthito 'lpavid api pratipadyaḥ sūktim |
śrībhāskaroditapitāmahatulyakarma-
vṛttiṃ vidhātum amalām aham udyato 'smi ||

See manuscript 61.

An anonymous vivṛti on Bhāskara's *Karaṇakutūhala* which mentions Śaka 1464 and 1466 = A.D. 1542 and 1544. The first verse is:

śaṇkaracaraṇasarojaṃ
praṇamya bhaktyā yathāmati mayeyam |
karaṇakutūhalavivṛttir
vidhīyate bālabodhāya ||

See manuscript 62.

The **BRAHMATULYODĀHARAṆA**, a commentary on Bhāskara's *Karaṇakutūhala* composed by Viśvanātha in part in 1612 at Kāśī, though there are examples dated as late as 1644 in at least some manuscripts. *CESS* A5, 669a–669b. The first verse is:

praṇamya vaghnahartāraṃ mandabodhāya samprati |
brahmatulyasya gaṇitodāhṛtiṃ prakaromy aham ||

Manuscripts:

64. Khasmohor 5580. Ff. 1–34. 11 × 27 cm. 11 lines. With some marginal notes. Lemmata only from the mūla. Incomplete.

F. 34v ends in the ṭīkā on 5, 6–7 (p. 68, 26–29): tātkālikaś caṃdraḥ 8|5|17|8| pātaḥ 9|18|46|52| sapātacaṃdraḥ 5|24|4|0 aṃgulādyaḥ.

65. Puṇḍarīka jyotiṣa 39. Ff. A and 1–21, ⟨22⟩, and 23–77. 9.5 × 25 cm. 7 lines. Lemmata only from mūla. Incomplete. Formerly property of Viśveśvara, the son of Rāmeśvara Puṇḍarika.

Ff. Av–1r. Blank.

Colophon of adhyāya 9 (p. 114) on f. 77: iti śrīdivākaradaivajñātmajaviśvanāthadaivajñaviracite
brahmtulyasyodāharaṇe pātādhikārasyodāharaṇaṃ.

After this is written: śrīviśvanāthena divākarasya sutena kāśyāṃ racitā samāptā gobhidhagrāmanivāsineyam udāhṛtiḥ kheṭakutūhalasya vedāgnitithyu1534nmitaśāke saṃvatsare durmatināmadheye pauṣe site rudratithau jñavāre samāptim āgād gaṇitaṃ samastaṃ.

Ownership note on f. A:

rāmeśvaratanūjasya viśveśvaravipaścitaḥ ||
brahmatulyodāharaṇapustakaṃ varttate tv idam || 1 ||

The **GAṆAKAKUMUDAKAUMUDĪ**, a commentary on Bhāskara's *Karaṇakutūhala* composed by Sumatiharṣa Gaṇi at Nikaṣāpurī in the Vindhyādri in *ca.* 1620. *CESS* A6. Edited by Satyendra Miśra, Vārāṇasī 1991. The first verse in the manuscripts, but not in the edition, is:

śambhuṃ svayambhuvam ahaṃ praṇipatya pūrvaṃ
nābhyudvahaṃ vigatakarmarajovitānam |
śvo(?) dharmabhūruhadhano dhṛtavān subhavyaḥ
kṣetre 'tra bodhivapanāya maho 'kṣalakṣma ||

Manuscripts:

66. Khasmohor 5576. Ff. 1–56. 11 × 26 cm. 12 lines. With some marginalia. Incomplete (adhyāya 11 omitted). Copied on Friday 25 March 1737.

Colophon on f. 56v: ity āṃcalikamahopādhyāyaśrīudayarājagaṇīnāṃ śiṣyopādhyāyaśrīharṣaratnagaṇīnāṃ śiṣyapaṃḍitaśrīsumatiharṣagaṇiviracitāyāṃ karaṇakutūhalavṛttau gaṇakakumudakaumudīnāmnyāṃ grahaṇasaṃbhavādhikāro daśamaḥ || 10 || || samāpto yaṃ graṃthaḥ.

Post-colophon: saṃ 1794 varṣe caitrasudi 5 bhṛgau lipīkṛtaṃ.

67. Khasmohor 5429. Ff. 1–77. 10 × 26 cm. 11 lines. Incomplete (adhyāya 11 omitted). Copied on Thursday 7 September 1769. Formerly property of Sūryabhānu.

Colophon on f. 77v: ity āṃcalikamahopādhyāyaśrīudayarājagaṇīnāṃ śiṣyopādhyāyaśrīharṣaratnagaṇīnāṃ śiṣyapaṃḍitaśrīsumatiharṣagaṇiviracitāyāṃ karaṇakutūhalavṛttau gaṇakakumudakaumudīnāmnyāṃ grahaṇasaṃbhavādhikāro daśamaḥ || 10 || samāpto yaṃ graṃthaḥ.

Post-colophon: śrīsaṃvat 1826 varṣe bhādrapadamāse śukle pakṣe tithau saptamyāṃ gurau lipīkṛtam | lekhakapāṭhakayoḥ śubhaṃ bhūyāt.

Ownership note: sūryabhānusya pustakam idam.

After this is written:

adṛṣṭidoṣāt mativibhramā⟨c⟩ ca
yad arthihīnaṃ likhitaṃ mayātra
tat sādhum⟨u⟩khyair api sodhanīyaṃ
kopo na kāryyaḥ khalu lekhakāya || 1 ||

68. Khasmohor 5055. Ff. 1–23, 23b, and 24–29. 11 × 25.5 cm. 17–19 lines. Usually only lemmata from the mūla. Incomplete.

F. 29v ends in the ṭīkā on 8, 4–6 (p. 98, 8): ubhayor mārgragrahayor madhye yo maṃ.

69. Khasmohor 5255. Ff. 1–7. 10.5 × 21.5 cm. 12–15 lines. Incomplete (adhyāya 9).

F. 1 begins with the heading and udāharaṇa for adhyāya 9 (*cf.* p. 103): atha pātādhyāyo vyākhyāyate tatrādau pātaka lisyate | śake 1539 laukikakātīvadi 10 bhaume.

F. 7v ends, in line 10, in the ṭīkā on 9, 15 (p. 112, 12): ādīkṛtya gamya.

An unknown ṭīkā on Bhāskara's *Karaṇakutūhala*.

Manuscript:

70. Khasmohor 157(a). 1 f. 11 × 25.5 cm. 15 lines. Incomplete (end of adhyāya 4). Lemmata only from the mūla.

Recto begins with the lemma of 4, 21c (p. 61, 15): || mānāṃtarārddheti |

Recto ends with an abbreviated form of 4, 24 (p. 63, 21–22): iti karaṇakutūhale | vidagdhabuddhiballabhe | caṃdragrahaṇādhikāro caturthaḥ.

Verso. Blank.

The **SŪRYATULYA** composed by Dāmodara in 1417. *CESS* A3, 101a, and A4, 108a. The first verse is:

siddhiḥ syāt sarvakāmānām udaye 'rthapradānataḥ |
yasya viśvaikanetrasya tasmai śrībhāskṛte namaḥ ||

Manuscript:

71. Khasmohor 5198. Ff. 1–25. 12 × 26 cm. 9 lines.

Colophon on f. 25v: iti śrīgaṇakacakracūḍāmaṇisakalaśāstraviśārada śrīpadmanābhasyātmajena taccaraṇāravimdayugalaprasādāptavidyena śrīdāmodareṇa viracite sūryatulyābhidhāne karaṇe miśrakādhyāyo daśamaḥ samāptaḥ samāptam idaṃ sūryatulyābhidhaṃ nāma karaṇaṃ.

The **GRAHALĀGHAVA**, also entitled **SIDDHĀNTARAHASYA**, composed by Gaṇeśa at Nandigrāma in 1520. *CESS* A2, 94a–100a; A3, 27b–28a; A4, 72a–74a; and A5, 69b–72b. Edited with the ṭīkās of Mallāri and Viśvanātha by S. Dvivedin, Bombay 1925. The first verse is:

jyotiḥprabodhajananī pariśodhya cittaṃ
tatsūktakarmacaraṇair gahanārthapūrṇā |
svalpākṣarāpi ca tadaṃśakṛtair upāyair
vyaktīkṛtā jayati keśavavāk śrutiś ca ||

Manuscripts:

72. Khasmohor 4951. Ff. 1–31. 9.5 × 22 cm. 7/8 lines. Folia old and torn. With marginal notes. Copied by Vāsudeva for his son at Kāśī on Wednesday 27 January 1647.

Colophon on f. 31: iti grahalāghave paṃcāṃgānayanagrahaṇādhikāraḥ iti śrīsakalāgamācāryavaryaśrīkeśavasāṃvatsarātmajagaṇeśadaivajñaviracite siddhāṃtarahasye grahalāghavākhyaṃ samāptaṃ.

Post-colophon: śāke 1568 māghaśudipratipadā budhe pūrṇīkṛtam idaṃ pustakaṃ vāsudevena kāsyāṃ putrapaṭhanārthaṃ.

73. Khasmohor 5601. ff. A, α 1–16, β 17–30, and 1b (β ff. 17–30 are of a different paper, and f. 1b of a third type of paper). α and β two different

scribes. 11.5 × 27 cm. 7 lines. (f. 1b 11 × 25.5 cm. 9 lines.) Right sides of folia torn. With many marginalia. Incomplete (upasaṃhārādhikāra omitted). Copied by Harikṛṣṇa on *ca.* 22 September 1653. Formerly property of Senāpati.

Ff. Av–1r. Blank.

Colophon on f. 30: iti paṃcāṃgasiddhiḥ.

After this is written:

> samatalamastakaparidhir
> vaṃisi⟨d⟩dho daṃtidaṃtajaḥ śaṃkuḥ |
> ta⟨c⟩chāyātaḥ proktaṃ
> jñānaṃ dikdeśakālānām || 1 ||

Post-colophon: harikṛṣṇenālekhi saṃvat 1710 aśvine māsi śukle 10.

F. 30v. Blank.

Ownership note on f. A: pustakaṃ śrīsenāpatīnāṃ maulyena gṛhitaṃ.

F. 1b contains the beginning of a vyākhyā (Raghurāma's? *CESS* A5, 380b–382b) on Raghurāma's or Mahādeva's (*CESS* A4, 379b–381a, and A5, 290a–290b) *Kālanirṇayasiddhānta*.

śrīgaṇeśāya namaḥ ||

> praṇamyaikaradaṃ devaṃ śāradāṃ gurum eva ca |
> kālanirṇayasiddhāṃtaṃ vyākurve viśadoktitaḥ || 1 ||
> prājñair ādyair hṛdyapadyair nibaṃdhāḥ
> klptā nānāvedaśāstrārthavidbhiḥ |
> bhāsvatsv eteṣv arthasaṃdarśanārthaṃ
> dīpākārā me kṛtir durjjarāsmāt || 2 ||
> jānaṃti ye bhūri subhūrividyā
> teṣāṃ nimittaṃ na mamaiṣa yatnaḥ |
> utpatsyate matsadṛśo pi kaścit
> kālo hy anaṃto vipulā dharitrī || 3 ||

atha graṃthakarttā svābhīṣṭaṃ pratijānīte ||

> śivaṃ kālatanuṃ natvā vipaścijjanatuṣṭaye |
> kālanirṇayasiddhāṃtaṃ vakṣye graṃthānusārataḥ || 1 ||

ahaṃ kālanirṇayasiddhāṃtaṃ vakṣya ity anvayaḥ kālaḥ parāparavyatikarayaugapadyavirakṣiprapratyayaliṃgaḥ | teṣāṃ viṣayeṣu pūrvapratyayavilakṣaṇānām utpattāv anyanimittāsaṃbhavād yad atra nimittaṃ sa kāla iti sarva.

74. Puṇḍarīka jyotiṣa 37. Ff. 1–19. 12 × 27.5 cm. 10 lines. Edges of folia torn. With many marginal notes. Copied on Friday 24 November 1704. Formerly property of Vaijanātha Jo⟨śī⟩.

F. 19v ends in upasaṃhārādhikāra 3d (p. 367, 18): śako yadi.

Upasaṃhārādhikāra 4–5 are written by a second scribe in the upper margin; colophon: (iti svapraśaṃsā).

The first scribe continues:

dviṣṭ⟨h⟩o ⟨'⟩yaṃ grahalāghavadyunicaya[ḥ]ś cakrāhataiḥ ṣaṭśaraiḥ 56
ṣaṭghasrai26ś ca yutaḥ savāṇatapan⟨aḥ⟩125 seṣu5ś ca khāṃgāgnibhi360⟨ḥ |⟩
khājyāśai30r vihṛtaḥ phale guṇayama23[a]ghne cakranichrā(?)5kṣakho-
pete dvitri32[b]yute nagai7r[c] varitake hastiḥ samāsa⟨+⟩ pau || 4 ||
dviṣṭ⟨h⟩o 2 dyupiṃḍo radabhū132yuto ⟨'⟩rka12yuk
khāṃgā360gnihṛt khāgni30hṛtaḥ phale yute ||
vedai4r guṇai3s tridvi2⟨3⟩hate ⟨'⟩dri7śeṣite
⟨'⟩rkād brahmatulye karaṇe ⟨'⟩bdamāsapau || 5 ||
vayo ⟨'⟩bdās tri3taṣṭā daśaghnā gatāyuḥ
samā māsaghasraikyayuktā samābhyaḥ ||
dināṃ15śānvitā janmatithyanvitās te
kharāmai[ḥ]30ś ca taṣṭā tithiḥ syād daśādau || 6 ||

a. 3s2. b. 2s3. c. nago7r.

Colophon on f. 19v: samāpto yaṃ grahalāghavaḥ.

Post-colophon: saṃvat 1761 āgrahāyaṇike māse 9 bhṛguvāsare.

Ownership note on f. 1: idaṃ jo. vaijanāthasya grahalāghavapustakaṃ varttate.

75. Khasmohor 5319. Ff. ⟨1⟩,2–13, and ⟨14⟩. 15 × 27 cm. 11–13 lines. With some marginalia. Copied by Santoṣarāma on Thursday 5 July 1705.

Colophon on f. ⟨14⟩: iti śrīnaṃdigrāmasthadaivajñagaṇeśaviracitaṃ siddhāṃtarahasyaṃ grahalāghavaṃ saṃpūrṇaṃ.

Post-colophon: dvirasabhūcaṃdramite (read: bhūbhṛc°) varṣe śrāvaṇe māsi kṛṣṇe pakṣe 11 gurau saṃtoṣarāmeṇa likhitam idam.

Price on f. ⟨1⟩ and f. ⟨14v⟩: kīmati || ≡ ||ɟ.

76. Khasmohor 5011. Ff. 1–178. 12 × 26.5 cm. 9 lines. With the ṭīkā of Mallāri. Copied by Tulārāma for Mahārāja Jayasiṃha on Thursday 5 September 1706 from a manuscript copied by Viṭṭhala Dīkṣita in 1625.

Colophon on f. 178: iti śrīmatgaṇakacūḍāmaṇidivākaradaivajñasūtamallāridaivajñaviracitāyā grahalāghavaṭīkāyāṃ graṃthasamāptyalaṃkāravyākhyānaṃ samāptaṃ.

After this follow Mallāri's two final verses (p. 370, 5–12).

Post-colophon on ff. 178–178v:

bāṇonā⟨c⟩ chakataḥ kurāma31vihṛtān mūlaṃ hi māsaḥ sa yuk
[f. 178v] bāṇair bhaṃ ca daśonitaṃ dinamitis tasyā dalaṃ syāt tithiḥ ⟨|⟩

pakṣaḥ syāt tithisaṃmito 'khilayutiḥ saptābdhitithyunmitā
śrīmadviṭṭhaladīkṣito ⟨'⟩likhad[a] imāṃ ṭīkāṃ praṇamyācyutaṃ |

a. khilad.

śrīmahārājādhirājaśrījayasiṃhajīdevājñayā likhitam idaṃ tulārāmeṇa || miti bhādrapadaśuddha 9 gurau.

Price on f. 1: kī ♉||=ɹ.

77. Khasmohor 5394. Ff. 1–54. 11 × 22 cm. 6 lines. With marginal notes. Incomplete (upasaṃhārādhikāra omitted). Copied by Harinārāyaṇa Praśnorā on Friday 22 June 1733.

Colophon on f. 54: iti śrīkeśasāṃvatsarātmajagaṇeśadaivajñaviracite grahalāghavākhye siddhāṃtarahasye paṃcāgānayanagrahaṇādhikāraḥ.

Post-colophon: saṃvat 1790 varṣe āsāḍhamāse kṛṣṇapakṣe tithau 8 śukravāre liṣitaṃ praśnorā harinārāyaṇena.

78. Khasmohor 5600. Ff. A and 1–50 (ff. 1–2, 37–40, 45–46, and 48–49 different paper and different scribe). 11 × 26.5 cm. 14 lines. With corrections by a different scribe, but not on the folia of different paper. With the ṭīkā of Viśvanātha. Copied by Paṇḍita Ḍūlīcandra, the pupil of Paṇḍita Sadāsukha Gaṇi, the pupil of Darśanasundara Gaṇi, the pupil of Kīrttiratna Sūri, at Mulatāna on Saturday 21 March 1741.

F. Av. Blank.

Colophon on f. 50: iti śrīdaivajñadivākarātmajaśrīviśvanāthaviracitaṃ siddhāṃtarahasyagrahalāghavākhyakaraṇodāharaṇaṃ samāptam astu (samāptam astu deleted).

After this is written: anuṣṭubhaḥ 1821 . . .

yādṛśaṃ pustakaṃ dṛṣṭvā tādṛśaṃ liṣitaṃ mayā
yadi śuddham aśuddhaṃ vā mama doṣo na dīyatāṃ 1
bhagnapṛṣṭikaṭigrīvā baddhadṛṣṭav adhomuṣaṃ
kaṣṭena liṣitaṃ śāstraṃ yatnena paripālayet 2
tailād rakṣej jalād rakṣed rakṣe sithalabaṃdhanāt
mūrṣahaste na dātavyaṃ evaṃ vadati pustakaṃ || 3 ||

Post-colophon: saṃvat 1798 śāke 1663 mitī caitrasudi 15 śanau śrīmulatānamadhye śrīkīrttiratnasūriśāṣāyāṃ upaśrīdarśanasuṃdaragaṇitaśiṣyapaṃ. sadasuṣagaṇitaśiṣyapaṃ. ḍūlīcaṃda lipīkṛtā.

F. 50v. Blank.

79. Puṇḍarīka jyotiṣa 269. Ff. 1–5 and f. ⟨A⟩. 10 × 21.5 cm. 9 lines. With marginal notes. Similar to manuscript 80. Incomplete. (tripraśnādhikāra 8–10). Copied on Monday 16 January 1826.

F. 1r begins, after the salutation, with tripraśnādhikāra 8a (p. 127, 1): vedeśāḥ 114.

On f. 3v ends the ṭīkā on tripraśnādhikāra 10 (p. 134, 18): iti pāṭhaḥ yuktaṃ.

Ff. 3v–5v continue with the lemmata of the verses on ff. 4–4v of manuscript 80 with a ṭīkā.

F. 5v ends:

mūlāvaśeṣakaṃ (saikaṃ) ṣaṣṭighnaṃ vikalānvitaṃ
dvisaṃguṇadviyuktena mūlenāptaṃ sphuṭībhavet

See the end of manuscript 245.

Date-formula: miti pauṣasudi | 11 || caṃdavāra saṃvat || 18 || 82.

On f. ⟨A⟩ a second scribe has written:

akṣa⟨c⟩chāyāvargatat⟨t⟩vāṃ25śayuktā
mārttaṃḍā syād aṅgulādyo kṣakarṇaḥ ||
tathākṣa
chāyeṣughnākṣabhāyāḥ kṛtidaśamalavonā yamāśāḥ palāṃśāḥ || 5 ||
atha vedābdhyabdhyū444naḥ kharasahṛtaḥ śako yanāṃśāḥ |

These lines are found at the end of f. 4 of manuscript 80.

80. Puṇḍarīka jyotiṣa 17. Ff, A, 1–4, and B. 10 × 21.5 cm. 9 lines. With marginal notes. With the ṭīkā of Viśvanātha. Incomplete (tripraśnādhikāra 8–10). Copied by Śyāmasundara for Dhaneśvarajī Pauṇḍarīka on Monday 6 March 1826.

Ff. A–Av. Blank.

F. 1r begins, after the salutation, with tripraśnādhikāra 8a (p. 127, 1): vedeśāḥ 114.

F. 3v ends in the ṭīkā on tripraśnādhikāra 10 (p. 134, 18): iti pāṭhaḥ || yuktaṃ.

There is no colophon.

Post-colophon: śrīḥ saṃvat 1882 miti māghakṛṣṇa 30 somavāsare likhitam idaṃ pustakaṃ śāmasuṃdareṇa śrīdhaneśvarajīpauṃḍarī(ka)paṭhanārthaṃ.

On ff. 4–4v are written the following verses:

+ + [f. 4]dos tribhonaṃ tribhorddh⟨v⟩aṃ viśeṣyaṃ
ra(sai)ś cakrato ⟨'⟩ṃkādikaṃ syād bhujonaṃ
tribhaṃ koṭir ekaikakaṃ tritribhaiḥ syāt
padaṃ sūryamaṃdoccam aṣṭādrayo ⟨'⟩ṃśā [bhavet] || 1 ||
meṣādige sāyanabhāgasūrye
dinārddhajā bhā palabhā bhavet sā ||
triṣṭ⟨h⟩ā hatā⟨ḥ⟩ syur daśabhi|10|r bhujaṃgai|8|r
ddigbhi|10|ś carārddhāni guṇo|3|ddhṛtāṃtyā || 2 ||
syāt sāyanoṣṇāṃśubhujarkṣasaṃkhy⟨ā⟩(?)
carārddhayogo lavabhogyaghātāt ||

khāgnyāptiyuktas tu caraṃ dhanarṇaṃ
tulājaṣaḍbhe tapane (')nyathāste || 3 ||
laṃkodayā vighaṭikā gajabhāni goṃka-
dasrās (t)ripakṣadahanā(ḥ) 278|299|323| kramagotkramasthāḥ ||
hīnānvitāś caradalaiḥ kramagotkramasthair
meṣādito ghaṭata utkramata(s t(v) i)me syuḥ || 4 ||
golau staḥ saumyayāmyau kriyaghaṭarasabhe khecare hāyane te
nakrāt kīṭāc ca ṣaḍbhe (')tha carapalayutonās tu pa(ṃ)ceṃdu15nāḍyaḥ ||
ghasrārddhaṃ golayo(ḥ) syā(t) tadayutakhaguṇāḥ 30| syān niśārddhaṃ
tathākṣa-
(c)chāyeṣughnākṣabhāyāḥ kṛtidaśamalavonāpamā(ṃ)śāḥ palāṃśāḥ || 5 ||

Verse 5 is Gaṇeśa, *Grahalāghava*, tripraśnādhikāra 6.

atha vedābdhyabdhyū444naḥ kharasahṛtaḥ śako (')nyanāṃśāḥ ||

[f. 4v]rasa6rttviṣū5dadhi4dvi2bhūmitair lavādiko 'pamaḥ ||
dināṃśa15labdhakhaṃḍakair bhavet sukhāptaye (')thavā || 1 ||
ṣaṭṣaḍiṣū5dadhi4dṛk2kku2bhir arddhaiḥ
kheṭabhujāṃśadināṃśamitaikyaṃ ||
śeṣahataiṣyadināṃśayutaṃ vā-
ṃśādyapamaḥ sukhasaṃvyavahṛtyai || 2

Verse 2 is Gaṇeśa, *Grahalāghava*, tripraśnādhikāra 12.

A second scribe has written in the margin of f. 4: saṃ 1888 śa 1753 ghāta 444 vākī 1309; and in the margin of f. 4v: bhāga 60 sāyana 21|49.
Ff. B–Bv. Blank.

81. Khasmohor 1086. Ff. 1–6. 16 × 28.5 cm. 9 lines. Folia torn. With some corrections. Incomplete.
F. 6v ends in pañcatārāspaṣṭīkaraṇa 17b (p. 111, 28): nagādri.
On f. 1 is written:

	1	2	3	4	5	6	7	8	9	10	11	12
0	41	81	117	150	178	199	212	212	195	155	89	0
0	25	47	68	84	98	106	108	102	79	66	36	0
0	63	126	186	246	302	354	402	440	461	443	326	0

82. Khasmohor 5086. Ff. 1–7. 14.5 × 31 cm. 14–16 lines. With the ṭīkā of Viśvanātha. Incomplete.

F. 7v ends in pañcatārāspaṣṭīkaraṇa 1b (p. 88, 3): dṛ.

83. Khasmohor 5122a. Ff. 1–2. 14 × 31 cm. 10 lines. With the ṭīkā of Viśvanātha. Incomplete.
F. 2v ends in ravicandraspaṣṭādhikāra 7d (p. 81, 19): bdhyabdhyū.

84. Khasmohor 5283. Ff. 1–76. 11 × 25.5 cm. Scribe α: ff. 1–45r, 8–10 lines; scribe β: ff. 45v–76v, 11–16 lines. Folia torn. With marginal notes. With the ṭīkā of Mallāri. Incomplete. Formerly property of Śiva Daivajña.
F. 76v ends in the ṭīkā on candragrahaṇādhikāra 12 (p. 192, 27): jñātavyā.
Ownership note on f. 1: śivadaivajñasyedaṃ

85. Khasmohor 5530. Ff. 1–12. 13.5 × 34 cm. 10–15 lines. With marginalia.
Colophon on f. 12: iti śrīnaṃdigrāmasthadaivajñagaṇeśaviracitaṃ siddhāṃtarahasyaṃ grahalāghavaṃ saṃpūrṇaṃ.
Price on f. 12v: kī |J.

The **GRAHALĀGHAVAṬĪKĀ**, a commentary on Gaṇeśa's *Grahalāghava* composed by Mallāri in *ca.* 1575/1600. *CESS* A4, 365b–367b, and A5, 285a. The first verse is:

> nāke nākeśamukhyāḥ suravaranivahāḥ santi ye 'nantasaṃkhyā
> nākhyām ākhyāty amīṣāṃ katham api ca manaḥpūrvakaṃ vāṅ madīyā |
> ekaṃ hityaikadantaṃ sakalasuraśiraḥsaṅghasaṅgharṣitāṅghriṃ
> śīghraṃ bhakteṣṭasiddhipradam iha hi suraṃ sādaraṃ taṃ namāmi ||

See manuscripts 76 and 84.

The **SIDDHĀNTARAHASYODĀHARAṆA**, a commentary on Gaṇeśa's *Grahalāghava* composed by Viśvanātha in about 1620. *CESS* A5, 669b–674b. The first verse is:

> jyotirvidguruṇā gaṇeśaguruṇā nirmathya śāstrāmbudhiṃ
> yac cakre grahalāghavaṃ vivaraṇaṃ kurve 'sya satprītaye |
> smṛtvā śambhusutaṃ divākarasutas tad viśvanāthaḥ kṛtī
> jāgrajjyautiṣavaryagokulaparitrāṇāya nārāyaṇaḥ ||

See manuscripts 79, 80, and 82. Additional manuscripts:

86. Khasmohor 5535. Ff. 1–47 and 54–58. 11 × 22.5 cm. 10 lines. Lemmata only from the mūla. Incomplete. Copied on *ca.* 16 April 1667.

F. 1 begins, after the salutation, with the heading to the pañcatārāspaṣṭīkaraṇādhikāra (p. 91, 1): athā bhaumādīnāṃ spaṣṭīkaraṇādhikāro vyākhyāyate.

F. 47v ends in the ṭīkā on grahayutyadhikāra 4 (p. 319, 26): natāṃśāḥ | te.

F. 54 begins in the ṭīkā on pañcāṅgacandragrahaṇānayanādhikāra 3 (p. 354, 19): 21|5|9.

Colophon on f. 58v: iti śrīdaivajñavaryyadivākarātmajavināthaviracitaṃ siddhāntarahasyodāharaṇam || samāptam.

After this comes a verse missing in the edition:

golagrāmanivāsino gurupadadvandvābjabhaktau ratasy-
āsīt tatra divākarasya tanayaḥ śrīviśvanāthāhvayaḥ |
tenedaṃ grahalāghavasya gaṇitaṃ spaṣṭīkṛtaṃ tad budhaiḥ
śodhyaṃ śuddham idaṃ tadā tu gaṇakaiḥ svānte sadā dhāryatāṃ || 1 ||

Post-colophon: saṃvat· 1724 vaiśākhaśukle akṣai tṛtīyāyāṃ liṣitam idaṃ pustakaṃ.

Price on f. 1: kī o||ȷ.

87. Khasmohor 5016. Ff. 1–114. 12 × 26 cm. 8/9 lines. With some marginalia. Sometimes the mūla is given in full, sometimes just lemmata are given. Copied by Tulārāma for Mahārājādhirāja Jayasiṃha on Friday 11 October 1706 (through the tripraśnādhikāra) and on Saturday 26 October 1706 (to the end).

Colophon on f. 41v (*cf.* p. 159, 3–4): iti śrīdivākaradaivajñātmajaviśvanāthadaivajñaviracite grahalāghave lagnādichāyāyaṃtrabhāgādiksādhananalikāvaṃdhādhikārasyodāharaṇaṃ samāptaṃ.

Post-colophon: saṃ 1763 varṣe kārttikakṛṣṇa 1 bhṛguvāsare śrīmanmahārājādhirājajaiyasiṃhadevajīkasyājñayā likhitam idaṃ tulārāmeṇa.

Colophon on f. 114: iti śrīdaivajñavaryadivākarātmajaviśvanāthaviracitaṃ siddhāṃtarahasyodāharaṇaṃ saṃpūrṇaṃ.

Post-colophon: saṃ 1763 kārttikaśukla 1 śanau lekhaḥ mahārājādhirājamahārājaśrījaiyasiṃhadevajīkasyājñayā likhitam idaṃ tulārāmeṇa.

88. Khasmohor 5355. Ff. 1–13. And Khasmohor 5362. Ff. 14–44. 11 × 25.5 cm. 17 lines. Jaina Nāgarī with occasional pūrvamātras. With some marginalia. Copied by ṛṣi Harṣā and muni Lakṣmīcandra, the pupil of Mahopādhyāya Jagaccandra, the pupil of Vā⟨caka⟩ Hīracandra, the pupil of Bhaṭṭāraka Jayacandra Sūri of the Pārśvacandrasūrigaccha, at Yovanera (Yodhapura?) on *ca.* 2 April 1710 from the favor of N⟨e⟩minātha.

Colophon on f. 44v: iti śrīdaivajñavaryadivākarātmajaviśvanāthaviracitaṃ siddhāntarahasyodāharaṇaṃ samāptaṃ | iti śrīgrahalāghavasya vṛtti samāptā.

Post-colophon: saṃvat 1767 varṣe caitraśukla rākādine | śrīpārśvacaṃdrasūrigacche | bhaṭṭārakaśrījayacandrasūrayas tadaṃtevāsino vā° śrīhīracaṃdrā |

tacchiṣya mahopādhyāyaśrījagaccaṃdrās tadaṃtevāsī munilakṣmīcaṃdreṇa lipīkṛtā || ṛ harṣāsahitena | śrīyovaneramadhye śrīnaminathaprasādāt.

89. Khasmohor 5122b. Ff. 1–26. 14 × 31 cm. 14–16 lines. With some marginalia. Usually lemmata only from the mūla. Incomplete (ends with pātādhikāra). Copied at Jayapura on Wednesday 1 September 1756.

Colophon on f. 26: iti śrīdivākaradaivajñātmajaviśvanāthadaivajñaviracitā grahalāghavākhyasiddhāṃtarahasyodāhṛtiḥ samāptā.

Post-colophon: saṃvat 1813 śāke 1678 bhādrapadamāsi śuklasaptamyāṃ budhe samāptaṃ śrījayasiṃhanāmarājanagare daḥ.

90. Puṇḍarīka jyotiṣa 35. ff. 1–85 and 85b–100. 10 × 26 cm. 7/8 lines. With some marginalia. Copied in 1788/9. Formerly property of Viśveśvara, the son of Rāmeśvara Puṇḍarīka.

Colophon on f. 99v: iti śrīdaivajñavaryadivākarātmajaviśvanāthaviracitaṃ śiddhāṃtarahasyodāharaṇaṃ samāptaṃ.

After this is written:

golagrāmanivāsino gurupadadvaṃdvābjabhaktau ratasy-
āsīt tatra divākarasya tanayaḥ śrīviśvanāthāhvayaḥ ||
tenedaṃ grahalāghavasya gaṇitaṃ spaṣṭīkṛtaṃ tad budhaiḥ
śodhyāṃ siddham idaṃ tadā tu gaṇakaiḥ svāṃte sadā dhāryatāṃ || 1
[f. 100]āyus tena ravīravarddhatu sadā hemaṃtarātrir yathā
lokānāṃ priyadarśano bhava sadā hemaṃtasūryo yathā ||
śatrūṇāṃ raṇaduḥsaho bhava sadā hemaṇtatoyaṃ yathā
nāśaṃyāṃtu tavārayo khalu sadā hemaṃtabhaṃ yathā || 1 ||

Post-colophon: saṃvat 1845.

Ownership note on f. 1:

rāmeśvaratanūjasya viśveśvaravipaścitaḥ ||
grahalāghavodāharaṇapustakaṃ varttate śubham || 1 ||

91. Khasmohor 5284. Ff. 2–79. 11.5 × 26 cm. 9/10 lines. Folia torn. With some marginalia. F. 52 fills a lacuna left by the scribe, who numbered the folia after supplying it. Ususally lemmata only from the mūla. Incomplete.

F. 2 begins in Viśvanātha's introduction (p. 4, 3): taśāstravicārasāracaturo.

F. 79v ends in an example of an eclipse-computation which I have not located in the published version of Viśvanātha's commentary: atha caṃdrabiṃbabhūbhāsādhanam āha vitryaṃśeśā iti (*cf.* māsagaṇādhikāra 13c) ⟨|⟩ piṃḍanādyaṃtaraṃ 3 asya ṣaḍaṃśaḥ 0|30| anena vitryaṃśeśā 10|40| n⟨ā⟩ḍyādipiṃḍasya vidyamānatvāt yuktā jātaṃ caṃdrabiṃba⟨ṃ⟩ 11|10| atha bhūbhāsādhanaṃ ⟨|⟩ piṃḍanāḍyaṃtara⟨ṃ⟩ 3 tri3ghnaṃ 9⟨|⟩ asya paṃcamāṃśena 1|.

92. Khasmohor 5536. A. ff. 1–7; and B. ff. 1–4 and ⟨5⟩. 11 × 23.5 cm. 9 to 11 lines. Incomplete (excerpts).

A f. 1r begins, after the salutation, in the introduction (p. 4, 6): maṃgalaṃ valaṃ vasantatikalenāha, and continues in the ṭīkā on madhyamādhikāra (p. 4, 8): sā śrutir vedo jayati | kīdṛśīti ślokenāha.

Shortly on f. 1r the text jumps to the ṭīkā on madhyamādhikāra 4–5 (p. 11, 10): tatrādau udāharaṇakramo liṣyate.

A f. 7v ends with the colophon to nakṣatracchāyādikāra (p. 308, 8): iti nakṣatrachāyādhikārodāharaṇaṃ.

B f. 1r begins with the heading to the udayāstādhikāra (p. 257, 17): athodayāstādhikārodāharaṇaṃ.

B f. ⟨5⟩v ends in an unidentified computation: 9|24|16 anena rahitaḥ 11|22|47 |13 kṣepakena 10|7|8 yuto jā.

93. Puṇḍarīka jyotiṣa 34. Ff. 1–52 and ⟨53⟩. 11 × 26 cm. 9/10 lines. Lemmata only from the mūla. Incomplete. Copied by and formerly property of Gokula, the son of Śambhūnātha Puṇḍarīka.

F. ⟨53⟩ ends in the ṭīkā on māsagaṇādhikāra 19 (p. 240, 26): atha.

F. ⟨53⟩v. Blank.

Ownership and scribal note on f. 1:

> śaṃbhūnāthatanūjasya gokulasya vipaścitaḥ ||
> grahalāghavodāhṛter hi pustaṃ jñeyaṃ vudhaiś śubhaṃ || 1 ||

tenaiva likhitam ity api śeṣaḥ ||

An anonymous **GRAHALĀGHAVAVĀRTTIKA**, a commentary on Gaṇeśa's *Grahalāghava* using the horoscope of Mahārāja Jayasiṃha as its udāhraṇa; this is dated Saturday 6 kṛṣṇapakṣa of Mṛgaśira in Saṃ 1745, Śaka 1610 = 3 November 1688. The first thirteen verses are:

> śrīmadvidyāguruṃ natvā dhyātvā devīṃ sarasvatīm ||
> grahalāghavasūtroktavārttikaṃ racayāmy aham || 1 ||
> asti svastimatī ramyāmbāvatī sālakāvatī ||
> yaśasvatsāgaraḥ śaśvat tatsthaḥ svacchāśayo 'likhat || 2 ||
> sakalāgamasārasvajñānavijñātakauśalāt ||
> gaṇeśavākyatas tattvaṃ viśadīkaravāṇi tat || 3 ||
> ekapañcāśadabdasya yātāḥ syur viśvanāḍikāḥ ||
> palāni caikavedāś ca trivedā akṣarāṇi tu || 4 ||
> iti brāhmyaṃ yātam āyur gaṇitāgame brahmaṇo
> divase kalpadvayaṃ kalpe ca te smṛtāḥ || 5 ||
> caturdaśāpi manavas tatrāyaṃ saptamo manuḥ ||
> vaivasvatas tatra mahaikasaptatiyugāni vai || 6 ||
> caturbhis tu yugair ekaṃ mahāyugam udīritam ||

tricaturlakṣaviṃśāḍhyasāhasrais tu caturyugī || 7 ||
varṣair atyaṣṭilakṣāṣṭāviṃśādhikasahasrakaiḥ ||
kṛtaṃ kṛtayugaṃ proktaṃ prathamaṃ sukṛtābhidham || 8 ||
catuḥsahasrair hīnaṃ tu dvāparaṃ viśvalakṣakaiḥ ||
catuḥṣaṣṭisahasrais tu tretā varṣāṣṭalakṣakaiḥ || 9 ||
caturlakṣādhikair dantasahasrair abdakaiḥ kaliḥ ||
aṣṭāviṃśatime caiva varttamāne mahāyuge || 10 ||
caturthaṃ caraṇaṃ caiva kalir nāma niveditam ||
yudhiṣṭhiramahārājarājyāt pravarttitaḥ kaliḥ || 11 ||
tatra trayo 'mī saṃjātā bhūpālāḥ śakakārakāḥ ||
trayo 'grato bhaviṣyanti kalau ṣaṭ śākakārakāḥ || 12 ||
yudhiṣṭhiro vedakṛtābhrarāmair
varṣais tato vikramabhūmināthaḥ ||
tato 'pi pañcānalacandravarṣaiḥ
śrīśālisadvāhanaśākakarttā || 13 ||

Manuscript:

94. Puṇḍarīka jyotiṣa 38. Ff. 1–23. 11 × 25.5 cm. 11–13 lines. Many folia stuck together. Incomplete.

On f. 2 is written: saṃvat 1745 varṣe śāke 1610 pravarttamāne śrīmajjayasiṃhajīmahārājasya janma.

Jayasiṃha's horoscope is given on f. 2v for: laukikamṛgaśiramāse kṛṣṇapakṣe 6 puṇyatithau śanivāsare ghaṭikā 25 palāni 32.

Planets	Text (Sidereal)	Computation (3 November 1688) (Tropical)
Saturn	Libra	221°
Jupiter	Capricorn	324°
Mars	Libra	24°(!)+180°=204°
Sun	Scorpius	231°
Venus' śīghra	Scorpius	
Mercury's śīghra	Scorpius	
Moon	Cancer	248°(!)
Rāhu	Aries	23°

F. 23 ends with the computation of the longitude of Mercury: iti tātkālikabudhaḥ 7|24|17|26 idaṃ spaṣṭaḥ.

An unnamed karaṇa composed by Sūrya in about 1410 (an udāharaṇa is dated 12 śukla(?)pakṣa of Pauṣa in Saṃ 1465 = *ca.* 30 December 1408). The perserved sections are:

udayāstanirūpaṇa. Verses 1–19.
⟨miśrika⟩. Verses 5–21.
grahaṇavicāra. Verses 1–13.
grahayutinirūpaṇa. Verses 1–4.

The first verse is:

sānnidhyān mahasāṃ nidher diviṣado 'dṛśyatvam aste kadāpy
astaḥ so 'bhihitas tadantaravatāṃ vyaktiḥ smṛtaś codgamaḥ ||
tatrālpaṃ vrajatāṃ raver avanibhūr gurvarkajānāṃ sadā
prāgasto hy udayo 'paratra himagor bahvāgatiḥ sūryataḥ ||

The last two verses are:

śāke bāṇarasāmareśagaṇite pauṣasya pūrvārddhake
dvādaśyāṃ gaṇitāgatā yutir abhūd devejyaśanyoḥ purā ||
rātrau taddivase tayos tu vivaraṃ dṛṣṭaṃ caturbhir lavair
ākāśe gaṇitasya kā gatir aho jyotirvidaḥ procyatām || 3 ||
yaś candreṇa yutiṃ karoti svacaro yābhir ghaṭībhir niśi
spaṣṭas tatsamayodbhavas tu gaṇitāt taccandrayor antaram ||
tenaivāntarito bhaved iti muhuḥ saṃsādhayed antaraṃ
kheṭānām iti vakti sūryagaṇako nānyo 'sty upāyo laghuḥ || 4 ||

A conjunction of Saturn and Jupiter occurred on 1 January 1405.

Manuscript:

95. Khasmohor 5446. Ff. 1–4. 12 × 28 cm. 12/13 lines. Incomplete.
F. 4v ends in grahayutinirūpaṇa 4d: laghuḥ.
There is no colophon.

D. KOṢṬHAKAS

A ⟨**KHAṆḌAKHĀDYAKASĀRIṆĪ**⟩ based on the *Khaṇḍakhādyaka*, which had been composed by Brahmagupta at Bhillamāla in 665; the date and author of the sāriṇī are unknown. See *SATE*, pp. 175–176, and the *Grahacārasamuccaya* (ms. 144).

Manuscript:

96. Khasmohor 5174(k). 1f. 11 × 27 cm. Incomplete.
Recto. Table of the solar equation for 1 to 90 degrees; the maximum is 2;10,49° at an argument of 90°. A second number accompanies each entry, ranging from 2,15 at 1° to 0,19 at 90°. Labeled śrīmatkhaṃḍakhādyai.
Verso. Table of the mean motions of the Sun, the Moon, and the lunar anomaly for 1 to 16 days.

A **SŪRYASIDDHĀNTASĀRIṆĪ** based on the *Sūryasiddhānta*, which was composed in *ca.* 800; the date and the author of the *Sūryasiddhāntasāriṇī* are unknown.

Manuscript:

97. Khasmohor 5174(j). 1f. 8 × 17.5 cm.
Recto. A list of the rotations of the planets in a Mahāyuga:

Sū⟨rya⟩	4,320,000
Caṃ⟨dra⟩	57,753,336
Ca⟨ndra⟩ u⟨cca⟩	488,199
Ca⟨ndra⟩ pā⟨ta⟩	232,246
Maṃ⟨gala⟩	2,296,832
Vṛ⟨haspati⟩	364,220
Śa⟨ni⟩	146,568
bhājaka (civil days)	1,577,917,828

A table of the longitudes of the beginnings and ends of the 27 nakṣatras.
A verse. kalpādau 1,955,880,000.

ṣaḍguṇasya[a] grahasyādes triṃśadbhaktād gṛhādikam
athavā paṃcabhir bhaktād rāśiḥ syād daśabhir lavāḥ | 1 |

a. ṣaṭguṇasya.

At right angles is written the following table:

Maṃ⟨gala⟩	164|28
Bu⟨dha⟩	144|225|50
Vṛ⟨haspati⟩	130|14
Śu⟨kra⟩	163|183|24
Śa⟨ni⟩	115|27|

Below this is written: sūryasiddhāṃte bhagaṇāḥ ⟨|⟩ sū 4,320,000 ⟨|⟩ kudināḥ 1,577,917,828 śodhitāḥ ⟨|⟩ śeṣam 488,199 ⟨|⟩ etair vāṭikā kṛtāsti ⟨|⟩ vāṭikānīte ucce lavādi 1|34|16 dhanaṃ kriyate ⟨|⟩ tadā sūryasiddhāṃtajaṃ bhavati ⟨|⟩ pāṭhabhedena uccabhagaṇā⟨ḥ⟩ 488,21[1]6 ⟨|⟩ yadā vāṭikānīte ucce 4|42|48 dhanaṃ tad eva |

Verso. A table of the motions of the planets in 7 days.

Sūryadinagatayaḥ

sū	caṃ	maṃ	vu	vṛ	śu	śa	rā	u
0	3	0	0	0	0	0	0	0
6	2	3	28	[2]0	11	0	0	0
53	14	40	38	34	12	14	22	46
57	4	5	46	54	54	2	15	47
0	0	0	0	0	0	0	0	0

The corresponding mean daily motions are:

Sun	0;59,8,14,...°
Moon	13;10,34,51...°
Mars	0;31,26,25,...°
Mercury's śīghra	4;5,32,17...°
Jupiter	0;4,59,8,...°
Venus' śīghra	1;36,7,42,...°
Saturn	0;2,0,17,...°
Lunar node	⟨-⟩0;3,10,42,...°
Lunar apogee	0;6,41.

A table of the motion of the Moon's, Mercury's, and Venus' anomalies (kendras) for 1, 7, and 14 days.

caṃke	7	14	vukeṃ	7	14	śukeṃ	7	14
0	3	6	0	0	1	0	0	0
13	1	2	3	21	13	0	4	8
3	27	54	6	44	29	36	18	37
53	17	34	24	42[b]	38	59	56	53
33[a]	[1]4	[2]9	10	13	27	33	54	48
7	51	43	31	37	14	26	8	16

a. read 52. b. read 49.

A table of the beginnings of the 4 caraṇas of each of the 27 nakṣatras.

The **KARAṆAKUTŪHALASĀRIṆĪ** or **BRAHMATULYASĀRIṆĪ** composed by Nāgadatta on the basis of the *Karaṇakutūhala* composed by Bhāskara at Vijjaḍaviḍa in 1183. See *SATIUS*, pp. 36–37. The first verse is:

natvā vallabhanandanaṃ tadanu gopālāṃhripadmadvayaṃ
jñātvā śrīguruvākyato hy aharniśaṃ sadyuktim evādhunā ||

siddhānteṣu yathoktakhecaravidhiḥ suspaṣṭakoṣṭhair muhur
madhyaspaṣṭavibhāgato grahagaṇān kurve dinaughād aham ||

Manuscripts:

98. Khasmohor 5253(a). Ff. A, 1–5, and B. 11 × 26 cm. Copied by Tejabhāna on Thursday 11 May 1769, probably at Mulatāna.

Ff. A–Av. Blank.

F. 1. The instructional verses beginning natvā vallabhanandanaṃ.

Colophon: iti śrīkarṇakutūhalasāraṇyāḥ karṇamaṇyā mūlapadyāni.

F. 1v. Table 13. Precession.

Table of the longitudes of the apogees of the Sun, Mars, Mercury, Jupiter, Venus, and Saturn.

Table of the mean daily motions of the planets.

Table of the daśāntaras in ghaṭīs of the Sun, the Moon, the lunar apogee, and the lunar node at Mulatāna.

Table of the Rāmabījas:

ra	caṃ	u	pā	maṃ	bu	vṛ	śu	śa
0	0	0	0	1	11	3		
2	15	30	30	20	40	10		
+	−	+	−	+	+	−		

Table of the ahargaṇa since the epoch of the *Karaṇakutūhala*, 24 February 1183, for every 30 years from Śaka 1470 = A.D. 1548 (133,331) to Śaka 1740 = A.D. 1818 (231,935).

Table of the ahargaṇa from 1 to 30 years (only the ahargaṇas for 1 to 15 years filled in).

F. 2. Table of the mean motions of the lord of the year, the accumulated epact, the Moon, the lunar apogee, and the lunar node for 20-year intervals from Śaka 1665–1765 (= A.D. 1743–1843) and for 1 to 20 years. With this and the tables on ff. 2v–3v compare tables 1 to 9.

F. 2v. Tables of the mean motions of Mars, Mercury, Jupiter, Venus, and Saturn for the same intervals.

F. 3. Tables of the mean motions of the Sun, the Moon, the lunar apogee, and the lunar node for 1 to 12 months, 1 to 4 intervals of 6 days, and 1 to 5 days.

F. 3v. Tables of the mean motions of Mars, Mercury, Jupiter, Venus, and Saturn for the same intervals.

Ff. 4–4v. Tables 10, 11, 14, 16, 18, 20, and 22. Equations of the center. The argument is for 3,6,9,...90°, however, instead of 1,2,3,...90°, and the entries for increments to the velocities are omitted. The maxima differ from what is expected of a Brāhmapakṣa text:

Planet	Khasmohor 5253(a)	Brāhmapakṣa
Sun	2;11,0°	2;10,54°
Moon	5;2,41°	5;2,31°
Mars	11;31,48°	11;12,53°
Mercury	4;28°	6;3,38°
Jupiter	5;6°	5;15,47°
Venus	1;45°	1;31,50°
Saturn	7;40°	7;38,35°

F. 5. Tables 15, 17, 19, 21, and 23. Equations of the anomaly. The entries are for 6,12,18,...180° instead of 1,2,3,...180°, and the hypotenuse is omitted. Again the maxima are not canonical.

Planet	5253(a)	argument	Brāhmapakṣa	argument
Mars	40;16°	132°	41;17,59°	130°
Mercury	21;31°	108°	21;36,48°	110°
Jupiter	11;31°	102°	10;59,1°	100°
Venus	46;21°	138°	46;30,28°	140°
Saturn	6;22°	96°	6;10,7°	100°

F. 5v. Table of the apogees with the number of years in which their longitudes increase by 0;0,1°.

su	maṃ	bu	vṛ	śu	śa
2	4	7	5	2	7
17	10	10	21	19	26
17	2	28	21	51	37
10	32	7	48	57	33
8	16	9	4	7	86
1	1	1	1	1	1

Table of the mean daily motions of the planets; see f. 1v.

Table of the elongations from the Sun necessary for the occurrences of the stations.

maṃ	bu	vṛ	śu	śa	
5	4	4	5	3	
14	24	10	13	25	va⟨kra⟩
6	7	7	6	8	
16	6	20	17	15	mā⟨rga⟩

Colophon on f. 5v:

> ṣaḍdvyaṣṭendumite varṣe rādhe ṣaṣṭhyāṃ gurau sudi |
> tejabhānena lipteyaṃ samūlā sāraṇī mayā || 1 ||

Ff. B–Bv. Blank.

99. Khasmohor 5424(a)+5254(j). Ff. 1–15. 11 × 26 cm. Copied by Tejabhāna on Wednesday 26 April, 3, 24, or 31 May 1769.

F. 1. (5424(a)). The instructional verses beginning with natvā vallabhanandanaṃ.

Colophon: iti śrīkarṇakutūhale vidagdhabuddhivallabha spaṣṭādhikāram.

F. 1v. atha vrahmatulyoktagrahāṇāṃ spaṣṭīkaraṇe koṣṭakānusāreṇa sūtraṃ pratipādayati...

Table of the ahargaṇa from the *Karaṇakutūhala's* epoch, 24 February 1183, for Śaka 1470 = A.D. 1548 (133,331) to Śaka 1740 = A.D. 1818 (231,935) in 30-year periods; and for 1 to 30 years.

Table of the days the Sun remains in each zodiacal sign and its daily motion while in that sign.

F. 2. Table 1. Mean motions of the Sun. But for 1–33 20 year periods instead of 1–60 here and in the other tables.

F. 2v. Table 2. Mean motions of the Moon.

F. 3. Table 3. Mean motions of the lunar apogee.

F. 3v. Table 4. Mean motions of the lunar node.

F. 4. Table 5. Mean motions of Mars.

F. 4v. Table 6. Mean motions of Mercury.

F. 5. Table 7. Mean motions of Jupiter.

F. 5v. Table 8. Mean motions of Venus.

F. 6. Table 9. Mean motions of Saturn.

F. 6v. Table 10. Equation of the center of the Sun. The changes in the daily motion are omitted here and elsewhere.

F. 7. Table 11. Equation of the center of the Moon. The maximum is 5;2,30° instead of 5;2,31°.

F. 7v. Table 14. Equation of the center of Mars.

F. 8. Table 16. Equation of the center of Mercury.

F. 8v. Table 18. Equation of the center of Jupiter.

F. 9. Table 20. Equation of the center of Venus.

F. 9v. Table 22. Equation of the center of Saturn. The maximum is 7;38,36° instead of 7;38,35°.

Ff. 10–10v. Table 15. Equation of the anomaly of Mars with the hypotenuse (the differences are omitted here and elsewhere). The maximum is 41;17,52° instead of 41;17,59°.

Ff. 11–11v. Table 17. Equation of the anomaly of Mercury. The maximum is 21;36,58° instead of 21;36,48°.

Ff. 12–12v. Table 19. Equation of the anomaly of Jupiter. The maximum is 10;58,23° instead of 10;59,1°.

Ff. 13–13v. Table 21. Equation of the anomaly of Venus.

Ff. 14–14v. Table 23. Equation of the anomaly of Saturn.

F. 15. Table 12. Just the solar declination and its differences; the luna latitude is omitted.

Colophon on f. 15:

> rasadvyaṣṭeṃdumāne ⟨'⟩bde rādhe kṛṣṇe tithāv aheḥ |
> lipteyaṃ tejabhānena samūlā sāraṇī budhe || 1 ||

F. 15v. Blank.

100. Khasmohor 5052. Ff. 1–12. 10.5 × 26.5 cm. Incomplete.

F. 1 A table of a function related to the length of a tropical year and week-day numbers for periods of 57 years from Śaka 1105 = A.D. 1183 to Śaka 2198 = A.D. 2276, and for 1 to 57 years. The entry for 57 years is 5,46,59; 6,5;14,48 × 57 = 5,46,59;3,36. The table continues for every pakṣa from amā of Caitra.

The mean motions are given as in the *Makaranda*, but with different parameters. The tables called vāṭikā give the mean motions for every 10 ghaṭikās so that the first entry multiplied by 6 is the mean daily motion.

F. 1v. Sun. 0;9,51,21,42,5,31,9° × 6 =0;59,8,10,12,33,6,54°. The Rāmabīja is +0;2°, the deśāntara +0;0,22°.

F. 2. Moon. 2;11,45,48,17,11,10° × 6 = 13;10,34,52,31,43,7,0°.

F. 2v. Lunar apogee. 0;1,6,48,58,26,25,27° × 6 = 0;6,40,53,50,38,32,42°.

F. 3. Lunar node. ⟨-⟩0;0,31,48,4,20,19,30° × 6 = ⟨-⟩0;3,10,48,26,1,57,0°.

F. 3v. Mars. 0;5,14,24,41,37,34,21° × 6 = 0;31,26,28,9,45,26,6°.

F. 4. Mercury. 0;40,55,23,30,15,7,20° × 6 = 4;5,32,21,1,30,44,0°.

F. 4v. Jupiter. 0;0;49,51,28,58,22,13° × 6 = 0;4,59,8,53,50,13,18°.

F. 5. Venus. 0;16,1,17,18,29,19,40° × 6 = 1;36,7,43,50,55,58,0°.

F. 5v. Saturn. 0;0;20,3,50,35,48,23° × 6 = 0;2,0,23,3,34,50,18°.

F. 6–6v. Table 10. Equation of the center of the Sun. With the change in daily motion.

F. 6v–7. Table 11. Equation of the center of the Moon. With the change in daily motion.

F. 7–7v. Table 14. Equation of the center of Mars for 1° to 180° by itself as elsewhere. The maximum equation is 11;13° at 91°.

F. 7v–8. Table 15. Equation of the anomaly of Mars by itself as elsewhere. The maximum is 41;17° at 125°–130°.

F. 8v. Table 16. Equation of the center of Mercury. The maximum is 4;27°.

F. 8v–9. Table 17. Equation of the anomaly of Mercury. The maximum is 21;31° at 109°–113°.

F. 9v. Table 18. Equation of the center of Jupiter. The maximum is 5;16°.

F. 9v–10. Table 19. Equation of the anomaly of Jupiter. The maximum is 10;59° at 100°.

F. 10v. Table 20. Equation of the center of Venus. The maximum is 1;26°.

F. 10v–11. Table 21. Equation of the anomaly of Venus. The maximum is 46;35° at 136°.

F.11v. Table 22. Equation of the center of Saturn. The maximum is 8;0°.

F. 11v–12. Table 23. Equation of the anomaly of Saturn. The maximum is 5;35° at 92°–97°.

F. 12v. 7 1/3 lines of prose in bhāṣā beginning: sāraṇī kā madhyama spaṣṭa karaṇa kī vidhi karṇakutūhale.

101. Khasmohor 5424(b). Ff. 1–13. 11 × 26 cm. Incomplete.

F. 1. Table 1. Mean motion of the Sun.
F. 1v. Table 2. Mean motion of the Moon.
F. 1v–2. Table 3. Mean motion of the lunar apogee.
F. 2–2v. Table 4. Mean motion of the lunar node.
F. 2v–3. Table 5. Mean motion of Mars.
F. 3–3v. Table 6. Mean motion of Mercury.
F. 3v–4. Table 7. Mean motion of Jupiter.
F. 4–4v. Table 8. Mean motion of Venus.
F. 4v–5. Table 9. Mean motion of Saturn.
F. 5–5v. Table 10. Equation of the center of the Sun. With the change in velocity as elsewhere.
F. 5v–6. Table 11. Equation of the center of the Moon. The maximum is 5;2,30° instead of 5;2,31°.
F. 6–6v. Table 14. Equation of the center of Mars. The maximum is 11;6,21° instead of 11;12,53°.
F. 6v–7. Table 16. Equation of the center of Mercury.
F. 7–7v. Table 18. Equation of the center of Jupiter. The maximum is 4;15,47° instead of 5;15,47°.
F. 7v. Table 20. Equation of the center of Venus.
F. 8. Table 22. Equation of the center of Saturn. The maximum is 7;38,36° instead of 7;38,35°.
F. 8–8v. Table 13. Precession for 1 to 66 10-year periods.
F. 8v. Table of the apogees of the Sun and the five planets.

ra	maṃ	bu	bṛ	śu	śa
2	2[a]	7	5	2	7
18	8	15	12	28	28
0	30	0	30	0	0
0	0	0	0	0	0

a. read 4.

F. 8v. Table of deśāntaras in degrees. minutes, and seconds.

ra	caṃ	maṃ	bu	bṛ	śu	śa	u	pā
0	0	0	0	0	0	0	0	0
0	5	0	1	1	1	0	0	0
44	24	2	32	13	41	2	19	0

F. 8v. Table of bījas as in *The Rājamṛgāṅka of Bhojadeva*, Aligarh 1987, p. 59.

ra	caṃ	maṃ	bu	bṛ	śu	śa	u	pā
0	0	1	11	3	4	1	0	0
2	15	2	4	10	3	3	0	30
0	0	0	0	10	0	0	30	0
+	-	+	+	-	-	+	+	-

Ff. 8v–9v. Table 15. Equation of the anomaly of Mars with its hypotenuse as elsewhere. The maximum equation is 41;21,27° at 127°.

Ff. 9v–10v. Table 17. Equation of the anomaly of Mercury. The maximum is 21;37,40° at 109°.

Ff. 10v–11. Table 19. Equation of the anomaly of Jupiter.

Ff. 11v–12. Table 21. Equation of the anomaly of Venus. The maximum is 46;18,1° at 120°.

Ff. 12–13. Table 23. Equation of the anomaly of Saturn.

Ff. 13–13v. Table 12. Only the solar declination. The maximum is 1415 = 23;35° instead of 24°.

F. 13v. Incomplete table of the traikya for 1°–107°. Below it is written: atha rājamṛgāṅkoktāni traikyāni. Cf. *The Rājamṛgāṅka*, p. 61.

102. Khasmohor 5504(c). Ff. 17–23. 11 × 25.5 cm. Incomplete.

F. 17. End of table 19. Equation of the anomaly of Jupiter with its hypotenuse as elsewhere for 137° to 180°.

Ff. 17v–18. Table 20. Equation of the center of Venus with the differences and changes in velocity as elsewhere.

Ff. 18–20. Table 21. Equation of the anomaly of Venus.

Ff. 20v–21. Table 22. Equation of the center of Saturn.

Ff. 21v–23. Table 23. Equation of the anomaly of Saturn.

F. 23v. Blank.

103. Khasmohor 5504(d). Ff. 38–43. 11 × 26 cm. Incomplete.

Ff. 38–38v. Table 1. Mean motion of the Sun.

Ff. 38v–39. Table 2. Mean motion of the Moon.

Ff. 39–39v. Table 3. Mean motion of the lunar apogee.

Ff. 39v–40v. Table 4. Mean motion of the lunar node.

Ff. 40v–41. Table 5. Mean motion of Mars.

Ff. 41–41v. Table 6. Mean motion of Mercury.

Ff. 41v–42. Table 7. Mean motion of Jupiter.

Ff. 42–42v. Table 8. Mean motion of Venus.

Ff. 42v–43v. Table 9. Mean motion of Saturn.

Ff. 43v. Table of ⟨cālanas⟩ for 0 to 18 days.

Table of the āptasthānaka for 245 to 253 ⟨days⟩. The entry for 243 is 4 0;36,15 and that for 253 6 1;46,15.

Table of entries for 1/2, 1, 2, 3, 4, 5, 5 1/2, 6, 6 1/2, 7, 8, 9, 10, 11, and 0 ⟨months?⟩. The entry for 1/2 is 0 15;20,7; for 1 it is 1 0;40,15; for 11 it is 11;2;4,21; and for 0 it is 0 2;7,38.

The **MAHĀDEVĪ** composed by Mahādeva in 1316. See *CESS* A4, 374a–376b, and A5, 288a–289a. The instructions are in verse and entitled *Grahasiddhi*; the tables are described by O. Neugebauer and D. Pingree, "The *Mahādevī* of Mahādeva", *PAPS* 111, 1967, 69–92, and in *SATIUS*, pp. 37–39. The first verse of the *Grahasiddhi* is:

siddhiṃ karotīśajakendrabhorvī-
nādīn śivau kṣetrapavāggurūṃś ca |
cakreśvarārabdhanabhaścarāśu-
siddher mahādeva ṛṣīṃś ca natvā ||

See manuscripts 138 and 140. Other manuscripts:

104. Khasmohor 5353(a). Ff. 1–2. 11 × 25 cm. 18–20 lines. The *Grahasiddhi* with a *Bālāvabodha*. With marginal notes. Copied by Premarṣi at Medanarapura in 1674/5. The description of this manuscript is incomplete so that not all of the details are secure.

The ṭīkā begins: śrīmahādevīśāstrasya bālābavodhaṃ karomi.

The mūla begins: siddhiṃ karotīśajakeṃdrabhorvī.

The ṭīkā ends: spaṣṭādhikāras tu bodhabodhāna(?) bodhitaṃ balabodhom idaṃ samyak jagākhyena(?) jagajjanān(?)

Colophon: iti.

Post-colophon: saṃ 1731 medanarapure liṣitaṃ premarṣiṇā

105. Khasmohor 5597. Ff. 1–4. 11.5 × 27 cm. 9 lines. Copied by Tejabhānu on Thursday 30 January 1755. Incomplete (*Grahasiddhi* only).

Colophon on f. 4: iti śrīmahādevakṛtā grahasoddhi sāraṇyā sūtraṃ samāptam.

Post-colophon: saṃvat 1811 phālgunamāse kṛṣṇapakṣe tithau 3 guruvāre likhitataṃ tejabhānuḥ.

106. Khasmohor 5598. Ff. 1–42. 11.5 × 27 cm. 13 lines. With marginal additions. With the ṭīkā of Dhanarāja. Copied by Tejabhāna on Friday 13 February 1756.

Colophon on f. 42v: iti śrīmahādevyāḥ ṭīkā samāptā.

Post-colophon: saṃvat 1812 śakaḥ 1677 māghe māse śuklapakṣe tithau trayodaśyāṃ 13 śukre likhitā tejabhānena.

107. Khasmohor 5056. Ff. 1–6. 11 × 26 cm. 17/18 lines. Diamond-shaped space left blank in the middle of each page. The *Grahasiddhi*. Wiṭh the ṭīkā of Dhanarāja. With some marginalia and diagrams. Incomplete.

The mūla ends on f. 6v in 16a: gajā veda.

The ṭīkā ends on f. 6v: vrahmatulyoktaprakāreṇābdakarmmasaṃskṛtarāsyādimadhyacaṃdraḥ

4|12|46|40|31|52| aṃśādyo yaṃ dvādaśabhaktaḥ la.

108. Khasmohor 5200. Ff. 1–150. 11 × 25 cm. Tables.

Ff. 1–30v. Table 9. N=0 to 59.

Ff. 31–60v. Table 10. N=0 to 59.
Ff. 61–90v. Table 11. N=0 to 59.
Ff. 91–120v. Table 12. N=0 to 59.
Ff. 121–150v. Table 13. N=0 to 59.
Price on f. 1: kī 2j.

109. Khasmohor 5249(a). Ff. 1–44. 11 × 26.5 cm. Tables.
Ff. 1–30v. Table 9. N=0 to 59.
Ff. 31–44v. Table 10. N=0 to 27.

110. Khasmohor 5254(g). Ff. 1–10. 11.5 × 26.5 cm. Tables. F. 9 torn.
F. 1. Table 1a for 1 to 60.
F. 1v. Table 1b for 1 to 60.
F. 2. Table 2a for 1 to 60.
F. 2v. Table 2b for 1 to 60.
F. 3. Table 3a for 1 to 60.
F. 3v. Table 3b for 1 to 60.
F. 4. Table 4a for 1 to 60.
F. 4v. Table 4b for 1 to 60.
F. 5. Table 5a for 1 to 60.
F. 5v. Table 5b for 1 to 60.
F. 6. Table 6a for 1 to 60.
F. 6v. Table 6b for 1 to 60.
F. 7 Table 7a for 1 to 60.
F. 7v. Table 7b for 1 to 60.
F. 8. Variant of table 8a for 1 to 60.
F. 8v. Variant of table 8b for 0 to 60.
F. 9 Table 8b for 1 to 60; and table of the longitudes of Rāhu for 1 to 27 avadhis.
F. 9v. Table 8a for 1 to 60 (entry for 60 torn off).
F. 10. Variant of table 1a for 1 to 60.
F. 10v. Variant of table 1b for 0 to 59.

111. Khasmohor 5352. Ff. 1–4. 10 × 26 cm. 8 to 10 lines. Jaina Nāgarī. With marginal glosses. The *Grahasiddhi*.

Colophon on f. 4: iti śrīmahādevakṛte grahasiddhau mahādevī saṃpūrṇā.

F. 4v. Two verses in corrupt Sanskrit on the vighaṭikās in 7 days (verse 1) and 7 nights (verse 2).

Tables of the guṇakās (mean yearly motions) and kṣepakās (epoch longitudes) according to the Saurapakṣa, the Āryapakṣa, and the Dṛkpakṣa. The epoch is Śaka 1480 = A.D. 1558.

112. Khasmohor 5595. Ff. 1–76. 11 × 27 cm. Tables. With some marginal notes.

Ff. 1v–16. Table 9. N=0 to 59.
Ff. 16v–31. Table 10. N=0 to 59.
Ff. 31v–46. Table 11. N=0 to 59.
Ff. 46v–61. Table 12. N=0 to 59.
Ff. 61v–76. Table 13. N=0 to 59.
Ff. 76. Table of the true longitudes of the Sun for 1 to 27 avadhis.
Table of the longitudes of Rāhu for 1 to 27 avadhis.

The **MAHĀDEVĪDĪPIKĀ**, a commentary of the *Mahādevī* composed by Dhanarāja at Padmāvatī in 1635. *CESS* A3, 124a–124b; A4, 117b; and A5, 150a. The first two verses are:

vāsudevaṃ hariṃ natvā śrīguroḥ pādapuṣkaram ||
vāgdevīṃ tapanādīṃś ca herambaṃ bhuvaneśvarīm || 1 ||
mahādevoktasāraṇyā grahāṇāṃ vidadhāmahe ||
vṛttiṃ śāstrānusāreṇa daivajñānāṃ sukhāptaye || 2 ||

See manuscripts 106 and 107.

The **MAHĀDEVĪBĀLĀVABODHA**.

See manuscript 104.

The **MAKARANDA** composed by Makaranda at Kāśī in 1478. *CESS* A4, 341a–343a, and A5, 268a–268b; *SATIUS* 39b–46b; and *SATE* 92. Edited with parts of the ṭīkās of Divākara and Viśvanātha by Rāmajanma Miśra, Vārāṇasī 1982. The first verse is:

śrīsūryasiddhāntamatena samyag
viśvopakārāya guruprasādāt ||
tithyādipatraṃ vitanoti kāśyām
ānandakando makarandanāmā ||

Manuscripts:

113. Khasmohor 5174(h). 1 blank f.; I ff. 1–6; and II f. 1, f. A, ff. 2–6, f. B, and ff. 7–9, ⟨10⟩, and 11–14; and f. C. Ff Ar and Cr copied by a second scribe. 11.5 × 25 cm. Tables.

I f. 1v. Verse 1; table 1 for Śaka 1688 tp 1784 (A.D. 1766 to 1862); table 2 for 1 to 16 years; and table 3 for 0 to 26 pakṣas.

I ff. 2–2v. Table 4 for 0 to 59 horizontal and 0 to 54 vertical.

I f. 3. Table 5 for Śaka1688 to 1784 (A.D. 1766 to 1812); and table 6 for 1 to 24 years.

I f. 3v. Table 7 for 0 to 14 months.

I ff. 4–4v. Table 8 for 0 to 59 horizontal and 0 to 54 vertical.

I f. 5. Table 9 for Śaka1688 to 1784 (A.D. 1766 to 1862); and table 10 for 1 to 24 years.

I f. 5v. Table 11 for 0 to 14 months.

I ff. 6–6v. Table 12 for 0 to 59 horizonal and 0 to 54 vertical.

II f. 1. Table 17, parts 1–2, for ⟨Śaka⟩ 1628 to 1742 (A.D. 1706 to 1820), and for 1 to 57 years.

II f. 1v. Table 17, part 3, for 1 to 26 pakṣas.

II F. A. Repeat of II ff. 1–1v by a second scribe.

II f. Av. Table 13 for Śaka1640 to 1784 (A.D. 1718 to 1862); table 14 for 1 to 24 years; table 15 for 1 to 12 zodiacal signs; and table 16 for 1 to 27 nakṣatras.

II f. 2. Table 18 for 1 to 1,0.

II f. 2v. Table 19 for 1 to 1,0.

II f.3. Table 20 for 1 to 1,0.

II f. 3v. Table 21 for 1 to 1,0.

II f. 4. Table 22 for 1 to 1,0.

II f. 4v. Table 23 for 1 to 1,0.

II f. 5. Table 24 for 1 to 1,0.

II f. 5v. Table 25 for 1 to 1,0.

II f. 6. Table 26 for 1 to 1,0.

II f. 6v. Table 36 for 1, 13, 14, 15, 16, and 17 days.

II f. B. Table of the mean motion of Ketu to 9 sexagesimal places for 1 to 1,0.

II f. Bv. Blank.

II f. 7. Table 27 for 1° to 90°.

II f. 7v. Table 28 for 1° to 90°.

II ff. 8–8v. Table 29 for 1° to 180°.

II ff. 9–9v. Table 30 for 1° to 180°.

II ff. ⟨10–10v⟩. Table 31 for 1° to 180°.

II ff. 11–11v. Table 32 for 1° to 180°.

II ff. 12–12v. Table 33 for 1° to 180°.

II f. 13. Table of lunar latitude for 1° to 90° (maximum is 270' at 90°); and table of digits of lunar latitude measured in sixths for 1 to 30 units of 6° (maximum 90 at 15).

II f. 13v. Table of half-durations of a solar eclipse for 1 to 22 (maximum 4;38 ghaṭikās at 22); table 42 for 1 to 36 viṃśopakas; table of solar diameters in digits and of solar velocities in minutes and seconds for 1 to 12 saṅkrāntis; table 37; and the skeleton of a table of the elongations from the Sun for the Greek-letter phenomena according to the Brāhmapakṣa.

II f. 14. Table of solar declinations for 1° to 90° (maximum 1440′ at 90°).

II f. 14v. Blank.

II f. C. 11 verses written by a second scribe. It begins: atha caṃdragrahaṇaṃ |

japadānādike proktaṃ yasmin jñaiḥ śreyam uttamam |
iṃdubhāskarayor vacmi tadgrahaṃ sacamatkṛtim | 1 |

Colophon: iti caṃdragrahaṇaṃ ||
9 verses, written by the same second scribe. It begins:

dor śāṃtakāle tribhahīnalagnaṃ
dvidhāyanāṃśaiḥ sahitaṃ vidheyam
tadbāhubhāgāḥ śaracaṃdrabhaktās
tallabdhakoṣṭe 'pamabhāgasaṃjñā | 1 |

Colophon: iti makaraṃdavivṛttau sūryagrahaṇaṃ.
II f. Cv. Blank.

114. Khasmohor 5174(g). 1 f. 20.5 × 10.5 cm.
F. 1. Table 52 of the *Makaranda* for lunar visibility. Horizontally the entries are for Rāhu in 1 to 12 zodiacal signs, vertically for the Sun in 1 to 12 zodiacal signs. Below this is written:

yāvaṃtyo ghaṭikā amābhinnā⟨ḥ⟩ kāryā⟨ś⟩ ca ṣaṣṭitaḥ |
dinamānaghaṭīyojyā yā⟨ḥ⟩ sthitā⟨ḥ⟩ śeṣaṣaṣṭitaḥ | 1 |
adhikā[a] cet samāyātā rāhuravyoś ca cakrataḥ |
tadā saṃdṛśyate caṃdra[b] ūnenaiva pradṛśyate | 2 |
iti caṃdradarśanaślokau[c].

a. adhiko. b. caṃdram. c. °ślokai.

On the verso is written:

amānāḍyūnaśūnyarttuyutā[a] dinamitir yadi |
ravirāhughaṭībhyo ⟨'⟩lpā tadā caṃdro na dṛśyate | 1 |

a. ° nāḍyūnasūnya°.

The **MAKARANDAPADDHATIKĀRIKĀḤ**, a commentary on Makaranda's *Makaranda* composed by Harikarṇa in 1610. *CESS* A6. The first verse is:

natvā vāgīśapādābjaṃ harikarṇo mahāmatiḥ ||
makarandakṛtau ślokāṃs tanute bālabodhakān ||

Manuscript:

115. Khasmohor 5496. Ff. 1–2. 11.5 × 25 cm. 14 lines. With some marginalia.
Colophon on f. 2v: iti makaraṃdapaddhatikārikāḥ.
After this the original scribe has written 7 verses:

maṃdāṃkaraṃdraṃ phalasādhane yad
gatiṃ svakīyāṃ guṇayec ca tena ⟨|⟩
vibhajya ṣaṣṭyā kalikādikaṃ tad
dhanarṇakaṃ madhyagatau vidheyam | 1 |
kulīranakrādigate svakeṃdre
maṃdasphuṭā sā gaditā purāṇaiḥ |
tadūnitā syāt svacaloccabhuktiḥ
svaśīghrakeṃdrasya gatiḥ kalādyā | 2 |
drā⟨k⟩keṃdrabhuktiḥ śīghrāṃtyaphalakoṣṭāṃtarāhatā ⟨|⟩
ṣaṣṭyāptā liptikādyaṃ svam asvam agre ⟨'⟩dhihīnake | 3 |
deyaṃ maṃdasphuṭe tau syāt sphuṭabhuktiḥ khacāriṇām |
ṛṇena bahunā śuddhā tadā vakragatir bhavet || 4 ||

vaidai rūpena dhṛtibhis tithyaharghaṭikā yutāḥ |
madhoḥ sitādito ⟨'⟩tho syuḥ sarpāsyāḥ parapatrajāḥ | 1 |
evaṃ krameṇa catvāri viśve dhiṣṇe ghaṭīṣu ca ⟨|⟩
rāmā yogeṣubhūvedās tadghaṭīṣu kramād yutāḥ | 2 |
evaṃ syāt sthūlapaṃcāṃgaṃ varttamānād bhaviṣyakam |
palais tu ghaṭikāṃtaḥ syād aṃtaraṃ spaṣṭatas tataḥ | 3 |
iti paṃcāṃgasiddhiḥ ||

The **MAKARANDAVIVARAṆA**, a commentary on Makaranda's *Makaranda* composed by Divākara in *ca.* 1620. *CESS* A3, 108b–109b; A4, 111a–111b; and A5, 141b–142a. Edited with Viśvanātha's *Makarandodāharaṇa* by Umāśaṅkara Śarman Miśra at Vārāṇasī in 1886. The first verse is:

prajñāṃ yataḥ prāpya kṛtapratijñaṃ
spardhāṃ vidhatte prasabhaṃ pratijñam |
ajño 'pi taṃ śrīśivanāmadheyaṃ
gurūpamaṃ svīyaguruṃ bhaje 'ham ||

Manuscripts:

116. Khasmohor 5299. Ff. 1–13. 12 × 26 cm. 9 lines. Includes the *Rājādinirṇaya*.
Colophons on f. 13v: iti śrīrājādinirṇaya samāptaḥ iti śrīmātaṃ makaraṃdavivaraṇaṃ samāptiṃ.
Price: kīmati |·J|

117. Puṇḍarīka jyotiṣa 29. 1 blank f.; and ff. 1–9. 10.5 × 19 cm. 13 lines. With marginal notes. Includes the *Rājādinirṇaya*. Copied by Mitrarāma Bhaṭa. Formerly property of Govinda Kāka.
Colophon on f. 9v: iti śrīsakalagaṇakasārvabhaumaśrīkṛṣṇadivajñasu-tanṛsiṃhasūnena

divākareṇa viracitaṃ makaraṃdavivaraṇaṃ samāptaṃ.

Post-colophon: liḥ bhaṭamitrarāma.

Ownership note by a third scribe on f. 9v: goviṃdakākopanāmaka.

Below the post-colophon and above the ownership note a second scribe has written:

> kaler arddhapramāṇena bhājayed bhogyakaṃ kaleḥ ||
> labdhaṃ tad dharmam ity āhur adharmo nyūnaviṃśatiḥ || 1 ||

The **MAKARANDAṬIPPAṆA** or **MAKARANDODĀHṚTI**, a commetary on Makaranda's *Makaranda* perhaps composed by Moreśvara, but in parts identical with Viśvanātha's *Makarandodāharaṇa*. Its epoch is 1622, but it uses an example for 1639. The first three verses are:

> yadīyapadapaṅkajaḥ smaraṇataḥ surendrādayas
> tadaiva kamalālayālayatayā vilāsānvitāḥ ||
> bhavanti bhavasambhavāvibhavanāśanāḥ sarvadā
> jagadvijayasaṃyutaḥ sa jayatīha moreśvarāḥ || 1 ||
> dhyātvā viśveśvaraṃ devaṃ
> natvā moreśvaraṃ tathā ||
> smṛtvā viṣṇupadāmbhojaṃ
> tato labdhāvabodhakam || 2 ||
> kurve 'haṃ makarandasya ṭippaṇaṃ chātratuṣṭaye ||
> durbodhaṃ mandabuddhīnāṃ vāsanāhīnacetasam || 3 ||

Manuscript:

118. Khasmohor 5197. Ff. 1–32. 12 × 26.5 cm. 8/9 lines. With some marginal notes. Copied by Tulārāma at the command of Mahārājādhirāja Jayasiṃha on Sunday 8 December 1706.

Colophon on f. 32: iti makaraṃdodāhṛtiḥ.

Post-colophon: saṃ 1763 varṣe mārgaśīrṣakṛṣṇa 30 ravau śrīmanmahārājādhirājajīśrījayasiṃhadevajīkasyājñayā

likhitam idaṃ tulārāmeṇa.

Price on f. 32v: kī ||=ȷ.

The **MAKARANDODĀHARAṆA**, a commentary on Makaranda's *Makaranda* composed by Viśvanātha at Kāśī in about 1634. *CESS* A5, 675b–676b. The first verse is:

> natvā gajānanaṃ devaṃ viśvanāthaḥ karoty asau ||
> uddāharaṇam uddānaṃ makarandasya yatnataḥ ||

Manuscripts:

119. Khasmohor 5378. Ff. 1–8. 11 × 25 cm. 18 lines. With some marginal notes.

Colophon on f. 8v: iti makaraṃde ravigrahaṇaṃ samāptam.

After the colophon are several lines in bhāṣā.

120. Khasmohor 5506. F. A; I ff. 1–8; and II ff. 1–8 and 8b–15. 15 × 23.5 cm. 16–19 lines. With marginalia. Incomplete. F. A is from a different manuscript copied by a different scribe; the beginning of the lines are cut off.

F. A. 3 lines of text on astrological medicine, beginning: yaṃ || kiṃcid ratnakośābhiprāyam uktaṃ | tatra yeṣu nakṣatreṣu jvaritasya mṛtyur uktaḥ tathā yeṣu ca kṛchreṇa bahubhir dinair jvaramuktir; and ending: graṃthalāghavāya svalpāṃtaratvāt tathā ca jyotiṣārke.

F. Av. 15 lines on the same subject, beginning: maṃdārānyadine mṛdudhruvacarakṣiprair valakṣe sudhāyoge bheṣajam ācaret; and ending: anyatra anyeṣu nakṣatreṣu rugṇo navabhir dinair virug bhavati etaj jyotiṣārkā.

I f. 2v ends near the beginning of line 5: yāni deśāṃtarayojanāni bhavaṃti tāni sarvāṇi.

I f. 3 begins: sūryasya madhyamagatyā guṇitāni kāryāṇi. With these two sentences *cf.* p. 6, 13–14.

I f. 8v ends: evaṃ caitraśuklapaurṇimāsyāṃ śanau āśu ghaṭīpaleṣu 30|44 meṣasaṃkrāṃtipraveśaḥ | evaṃ vṛṣād iṣṭapi. See p. 29, 16–17.

II f. 1 begins: vahudhā yasmin dine prathamatithir bhavati tasminn eva dine meṣasaṃkramaṇaṃ bhavati. See p. 29, 13–15.

II f. 8v ends at the beginning of line 16: ṣaḍbhakte labdhāni phalāni | etāni pūrvasthāpitadinamāna. See p. 68, 19–p. 69, 1.

II f. 8b begins: spaṣṭīkaraṇaṃ | śukralatā 43|19|30|19 deśāṃtaraṃ |4|50|.

II f. 13r. Blank; f. 13v follows f. 12v.

II f. 15v ends, after iti rohiṇīcakraṃ ||: viṃśatyā guṇayed grāsaṃ svasvamānena bhājayet | viśvātmakagrahaṇaṃ | saṃva 1716 morvo māpyenātmediva(?).

Below this are a table of rājas and a rohiṇīcakra.

The **PĀTASĀRAṆĪ** composed by Gaṇeśa at Nandigrāma in 1522. *CESS* A2, 100b; A4, 74a; and A5, 72b. The first verse is:

vedābdhīndraviyukśakād guṇahatāt khāśvāptanāḍyūnitā
rudrāḥ sārdhajināś ca sāvayavake tattulyayoge gate ||
pātaḥ syād vyatipātavaidhṛta itīhetā yuter nāḍikāḥ
ṣaṣṭyāptāḥ svaghaṭīguṇāḥ sphuṭatarās tasmād apītaiṣyakaḥ ||

Manuscript:

121. Khasmohor 5517. Ff. 1–6. 12 × 25.5 cm. Tables. With some marginal notes.

Colophon on f. 6: iti śrīmatkeśavasāṃvatasarātmajagaṇeśadaivajñaviracitā pātasāraṇī samāptāḥ.

The **PĀTASĀRAṆĪVIVARAṆA**, a commentary on Gaṇeśa's *Pātasāraṇī* composed by Divākara in about 1630.

The first verse is:

śrīmacchivākhyaṃ gaṇitajñacakra-
cūḍāmaṇiṃ sajjanavṛndavandyam ||
vidur vido yaṃ dhiṣaṇena tulyaṃ
taṃ naumi nityaṃ dhiṣaṇāptihetoḥ ||

Manuscript:

122. Khasmohor 5518. Ff. 1–3. 12 × 25.5 cm. 9 lines. Copied on *ca.* 20 November 1705.

Colophon on f. 3v: iti pātasāraṇīvivaraṇam.

Post-colophon:

saptāśvirasabhūśāke pauṣakṛṣṇātmabhūtithau ||
alekhīdaṃ mayā tatra hetūr gurukṛpai mai ||

Price on f. 1: kī S|||.

The **BṚHATTITHICINTĀMAṆI** composed by Gaṇeśa at Nandigrāma in 1552. *CESS* A2, 104a–104b; A3, 28a; A4, 75a–75b; and A5, 73b; *SATIUS* 50b–51a; and *SATE* 101. The canons were edited with the ṭīkā of Viṣṇu by V. G. Āpaṭe in *ASS* 120, Poona 1942, 2nd part. The first verse is:

natvā brahmaharīśvareśvarasutāryārkādikheṭān dvijo
'hno 'rdhenābdadinādisiddhidam ahaṃ tithyādicintāmaṇim ||
kurve 'tyalpakṛtiṃ vidhāya bahulaṃ yatnaṃ gaṇeśaḥ kṛtī
pūrvābhyo 'ticamatkṛtiṃ tithikṛtiṃ paśyantu sujñā iha ||

The **SUBODHINĪ**, a commentary on Gaṇeśa's *Bṛhattithicintāmaṇi* composed by Viṣṇu in about 1610. *CESS* A5, 703b–704a. The first verse is:

yatpādāmbujadarśanāt paramanirdoṣasphuradrūpiṇī-
svasvājñānaghanāndhakāram anayā cetogṛhaṃ śudhyati ||
śuddhe cetasi cātmacintanam ato muktiḥ kim asyāḥ paraṃ
tasmāt taṃ gurum ātmarūpam aparaṃ nityaṃ namaskurmahe ||

Manuscript:

123. Puṇḍarīka jyotiṣa 31. Ff. 1–21. 11.5 × 23.5 cm. Lines per page not recorded. Only lemmata cited from the mūla. Copied on Sunday 15 November 1730. Formerly property of Jagannātha Samrāṭ, and later of Viśveśvara, the son of Rāmeśvara.

Colophon on f. 21: iti śrīsakalāgamācāryavaryadivākaradaivajñasutaviṣṇu-daivajñaviracitau

bṛhaccimtāmaṇivāsanābhāṣye sūkṣmarkṣadeśāmtarādisādhanādhyāyaḥ.

Post-colophon: saṃvat 1787 mitī mārgaśīrṣakṛṣṇa 2 rāvau pustakaṃ samāptaṃ.

Ownership note after the post-colophon: idaṃ pustakaṃ śrīsamrāṭjījagannāthasya.

Ownership note on f. 1:

> rāmeśvaratanūjasya viśveśvaravipaścitaḥ||
> bṛhaccimtāmaṇeḥ pustaṃ jagaj jānātu sarvadā |

The **SUBODHĀ** composed by Jayarāma at Alindrapūrī in 1557. *CESS* A3, 62b, and A4, 96a. The first verse is:

> praṇamya gopālapadāravindaṃ
> bhavaṃ bhavānīṃ gaṇanāyakaṃ ca ||
> śrīsūryapūrvān akhilān nabhaścarān
> vakṣe 'tisūkṣmāṃ sugamāṃ subodhām ||

Manuscript:

124. Khasmohor 5519. F. A; and ff. 1–21. 13 × 23.5 cm. F. A: 13 lines; ff. 1–21: tables.

Colophon on line 2 of f. Av: iti subodhāsāraṇī samāptāḥ.

Below this is the skeleton of a table.

Ff. 1–1v. Table of the tithidhruva, the tithikendra, and the tithivārādi for Śaka 1479 to 1696 (= A.D. 1557 to 1774) in steps of 31 years and for 1 to 31 years.

Ff. 2–2v. Table of the nakṣatradhruva, nakṣatrakendra, and nakṣatravārādi for Śaka 1479 to 1743 (= A.D. 1557 to 1821) in steps of 24 years and for 1 to 24 years.

Ff. 3–3v. Table of the yogadhruva, yogakendra, and yogavārādi for Śaka 1479 to 1743 (= A.D. 1557 to 1821) in steps of 24 years and for 1 to 24 years.

Ff. 4–4v. Table of the tithikendra and the tithivārādi with their cālakas for 1 to 38 avadhis.

Ff. 5–5v. Table of the same sort for nakṣatras.

Ff. 6–6v. Table of the same sort for yogas.

Ff. 7–11v. Table of tithiphalas.

Ff. 12–16v. Table of nakṣatraphalas
Ff. 17–21v. Table of yogaphalas
Price on f. 21v: kīmati || – |.

The **GRAHAKALPATARU** or **MAṆIPRADĪPA** composed by Raghunātha at Kāśī in 1565. *CESS* A5, 372a–372b. The first verse is:

natvā vāraṇarājavaktram amarair nirvighnatākāṃkṣibhir
juṣṭaṃ śaṃkaram īśitāram api ca brahmādinā kaukasām ||
saure kalpatarum pravacmi gaṇite śrīsomabhaṭṭātmajaḥ
kāśyāṃ śrīraghunāthabhaṭṭagaṇakaḥ prītyai sa bhūyān satām ||

Manuscripts:

125. Khasmohor 5249(b). Ff. 18–52; also numbered 1–35 in upper left margins of the versos. 11 × 25.5 cm. Tables. Copied by Tejobhānu on Thursday 29 September 1757. The mean motion tables are for 1,000 to 9,000 days, 10,000 to 90,000 days, 100,000 to 900,000 days, and 1,000,000 to 3,000,000 days.

F. 18. Mean motion table of the Sun; at the end spaces for 1,000,000 to 7,000,000 days, but only that for 1,000,000 days filled in. The mean motion in 1,000 days is 8^s 25;36,9,33°, in 1 day 0;59,8,10,10°.

F. 18v. Mean motion table of the Moon to 3,000,000 days. The mean motion in 1,000 days is 7^s 6;21,7,43°, in 1 day 13;10,34,52,4°.

F. 19. Mean motion table of the Moon's apogee to 200,000 days. The mean motion in 1,000 days is 3^s 21;22,55,11°, in 1 day 0;6,40,58,31°.

F. 19v. Mean motion table of the Moon's node to 100,000 days. The mean motion in 1,000 days is -1^s 22;59,8,37°, in 1 day -0;3,10,44,55°.

F. 20. Mean motion table of Mars to 3,000,000 days. The mean motion in 1,000 days is 5^s 14;1,9,46°, in 1 day 0;31,26,28,11°.

F. 20v. Mean motion table of Mercury to 300,000 days. The mean motion in 1,000 days is 4^s 12;19,4,58°, in 1 day 4;5,32,20,42°.

F. 21. Mean motion table of Jupiter to 2,000,000 days (spaces to 6,000,000 days). The mean motion in 1,000 days is 2^s 23;5,46,50°, in 1 day 0;4,59,8,49°.

F. 21v. Mean motion table of Venus to 1,000,000 days. The mean motion in 1,000 days is 5^s 12;8,47,1°, in 1 day 1;36,7,43,37°.

F. 22. Mean motion table of Saturn to 3,000,000 days. The mean motion in 1,000 days is 1^s 3;26,21,30°, in 1 day 0;2,0,22,53°.

F. 22v. Mean motion of the Sun for 1 to 9 days, 10 to 90 days, and 100 to 1,000 days; and for days 14, 15, and 16 and 354, 355, 365, 384, and 385.

F. 23. Mean motions of the Moon for the same numbers of days.

F. 23v. Mean motions of the Moon's apogee for the same number of days.

F. 24. Mean motions of the Moon's node for the same numbers of days.

F. 24v. Mean motions of Mars for the same numbers of days.

F. 25. Mean motions of Mercury for the same numbers of days.

F. 25v. Mean motions of Jupiter for the same numbers of days.

F. 26. Mean motions of Venus for the same numbers of days.

F. 26v. Mean motions of Saturn for the same numbers of days.

F. 27. Table of the weekdays on which the Sun enters each of the twelve zodiacal signs; the entry for 1 is 3;56,22.

Table of the weekdays on which the Sun enters each of the twenty-seven nakṣatras; the entry for 1 is 3;56,31,4.

F. 27v. A four-by-four square of unknown purport (not magic).

10	12	9	46
11	7	31	14
0	27	44	6
2	14	25	27

Ff. 28–30. Table of the true longitudes and daily moitons of the Moon for 1 to 248 days (nine anomalistic months). The entries for day 1 are 12;3° and 726′, for 248 days 27;44° and 723′.

The first set of tables of the equations of the planets are for 1 to 20 units of 9° of increases in the argument and a column 21 for 0° of argument; the functions tabulated are the śīghra and manda equations expressed in minutes and the hypotenuse expressed in units of which there are 60 in the radius of each deferent.

F. 30v. Table of the equations of Mars. The maximum śīghra equation is 2420′=40;20° at 13×9=117°; the maximum manda equation is 690′=11;30° at 9=81° and 10=90°; and the extreme hypotenuses are 99 and 20.

Table of the equations of Mercury. The maximum śīghra equation is 1290′= 21;30° at 11=99°; the maximum manda equation is 267′=4;27° at 9=81°; and the extreme hypotenuses are 82 and 37.

F. 31. Table of the equations of Jupiter. The maximum śīghra equation is 690′=11;30° at 10=90°; the maximum manda equation is 305′=5;5° at 9=81°; and the extreme hypotenuses are 72 and 48.

Table of the equations of Venus. The maximum śīghra equation is 2780′=46;20° at 14=126°; the maximum manda equation is 104′=1;44° at 9=81° and 10=90°; and the extreme hypotenuses are 103 and 16.

F. 31v. Table of the equations of Saturn. The maximum śīghra equation is 380′=6;20° at 9=81° and 10=90°; the maximum manda equation is 460′=7;40° at 9=81° and 10=90°; and the extreme hypotenuses are 66 and 54.

The second set of equation tables (actually true longitude tables) has as its double argument mean longitudes of the planets horizontally and the śīghra argument vertically. The mean longitudes, at intervals of 20°, are 0°; 10°; 30°; 50°; 70°; 90°; 110°; 130°; 150°; 170°; 210°; 230°; 250°; 270°; 290°; 310°; 330°; and 350°. The śīghra arguments, at intervals of 10°, are 0°; 10°; 20°; 30°; and so on to 350°. The entries are the true longitudes, with the śīghra equation computed from the śīghra argument, the manda from the distance of the mean longitude from that planet's apogee.

Ff. 32–36v. Table for Mars.

Ff. 37–40v. Table for Mercury.

Ff. 41–44v. Table for Jupiter.

Ff. 45–48v. Table for Venus.
Ff. 49–52. Table for Saturn.
F. 52v. Table of the beginnings of the 108 nakṣatracaraṇas.
Colophon: iti śrīraghunāthabhaṭṭaviracito grahakalpataruḥ samāptaḥ.
Post-colophon: saṃvat 1814 dvi āśvina vadi 2 gurau likhitaṃ tejobhānunā.

126. Khasmohor 5594. Ff. 1–8. 11 × 28 cm. 11 lines. Instructions only, with some tables of functions described in verse.

Colophon on f. 8v: iti śrīraghunāthabhaṭṭaviraciti grahakalpataruḥ samāptaḥ.

After this is written:

> śakaḥ pūrṇābhraśakrai1400rahita inahato māsayug[a] dvir bhagāśvī 27 |
> yuktonaś[b] caṃdraṣaṭkāṃ61śaviyug atha radā32prāptamāsair[c] yug ūrddh⟨v⟩aḥ |
> triṃśannighno digāḍhyo dvir iṣuśara55yutaḥ svāgnipūrṇādri[d]703bhāgair
> yuktono ⟨'⟩bdhyaṃga64labdhāvamaviyug aparo jīvapūrvo dyuvṛṃda⟨ḥ⟩ || 1 ||
> ravau rudrabāṇāḥ śarābdhi⟨ḥ⟩[e] khalokā
> vidhau śaṃbhubhūpākṛtiḥ[f] saptalokā⟨ḥ⟩ ||
> vidhūcce gajāḥ śakralokās trivarṇā
> nāgāṣṭāśvinaś cākṛtiḥ siddha pāte ⟨|| 2 ||⟩
> kuje ⟨'⟩rthā navākṣīṣubāṇā dhṛtiḥ syād
> buddhocce yamāsth⟨y⟩aṃkadasrā radā⟨ḥ⟩ syuḥ |
> gurau pūrṇam aṣṭeṃdava⟨ḥ⟩ śakrabhūpāḥ |
> site lokasūryā manur naṃdasaṃkhyāḥ |⟨| 3 ||⟩
> śanau vedasapteṣavaḥ pūrṇatulyāḥ
> kramāt kṣepakāś ceti rāśyādaya⟨ḥ⟩ syuḥ ||

a. māsayuk. b. yuktopaś. c. radā 32 prāptimāsair. d. khāgnipūrṇādri. e. śarābdhī. f. °bhūpākṣatiḥ.

Next to this is a table of the epoch mean longitudes; I present modern computations for 5 March 1478 in column 3.

Planet	Text (sidereal)	Computation (tropical)
Sun	11^s 5;45,30°	354°
Moon	11^s 16;22,37°	5°
Apogee	8^s 14;3,43°	
Node	8^s 28;22,24°	137°(!)
Mars	5^s 29;55,18°	209°
Mercury's śīghra	2^s 2;29,32°	
Jupiter	0^s 18;14,16°	27°
Venus's śīghra	3^s 12;14,9°	
Saturn	4^s 7;5,0°	150°

The **TITHICŪḌĀMAṆI** composed by Rāmacandra between 1560 and 1580. *CESS* A5, 479b–480a. The first verse is:

praṇamya puṣpadantākhyau śāradāṃ gaṇanāyakam ||
tithisiddhiṃ kāmadugdhāṃ karomy alpamatiḥ sphuṭām || 1 ||

This and several other of the 18 verses of the introduction are taken from the *Kāmadhenu* composed by Mahādeva in 1357. In order to distinguish the *Tithicūḍāmaṇi* I cite its second verse:

vighneśaṃ ca giraṃ guruṃ hariharaṃ brahmārkamukhyagrahān
natvā caiva guroḥ padābjayugalaṃ prāptaprasādaḥ sudhīḥ ||
śrīmadbhāskarakeśavādiracitān vīkṣyātha cūḍāmaṇiṃ
śiṣyaprārthanayā karomi sugamaṃ tithyādisiddhyai sphuṭaṃ || 2 ||

Manuscript:

127. Khasmohor 5570. F. 1. 15 × 27 cm. 13 lines. Instructions only, without tables.
Colophon on line 6 of f. 1v: iti śrīrāmacaṃdraviracitaṃ tithicūḍāmaṇi saṃpūrṇṇa.

The **KHEṬASIDDHI** composed by Dinakara in 1578. *CESS* A3, 104a–104b, and A5, 139a; *SATIUS* 53 (Anonymous of 1578); and *SATE* 101–112. The first verse is:

natvādau gaṇanāyakaṃ paśupatiṃ vāgdevatāṃ khecarān
sūryādīn kuladevatāṃ nijaguruṃ lakṣmīpatiṃ padmajam ||
kurve daivajidāṃ varo dinakaraḥ śrīkheṭasiddhiṃ mudā
svalpāṃ brāhmamatāśritāṃ sugahanāṃ vidvajjanānandinīm ||

Manuscript:

128. Khasmohor 5125. Ff. 1–3. 10.5 × 25 cm. 9/10 lines. Instructions only. Incomplete.
F. 3v ends in 31c–d (numbered 35 in manuscript): dasrāśvī22pramitād ardhā śuddhi syāt va.

The **CANDRĀRKĪ** composed by Dinakara at Bārejya in 1578. *CESS* A3, 102b–104a; A4, 109a; and A5, 138a–139a; *SATIUS* 51b–53a; and *SATE* 101. The first verse is:

sūryaṃ candraṃ sadguruṃ bhaktipūrvaṃ
natvā vakṣye sūryacandrodbhavaṃ ca ||
pattraṃ pañcāṅgābhidhaṃ buddhivṛddhyai
grāhyaṃ tajjñair yuktimat tan mayoktam ||

Manuscripts:

129. Khasmohor 5247. Ff. 1–2. 11 × 27 cm. 11 lines. Instructions only. With marginal notes. Copied on Friday 22 November 1754.
Colophon on f. 2v: iti śrīmaddinakarācāryaviracitā candrārkī saṃpūrṇaḥ.
Post-colophon: saṃvat 1811 mārgaśīrṣe śukle pakṣe tithau 8 śukre.

130. Khasmohor 5015(a). Ff. 1–8 and 8b–10. 12 × 28 cm. Tables.
Ff. 1v–4v. Table 1 of the *Candrārkī.*
Ff. 4v–7v. Table 2 of the *Candrārkī.*
Ff. 8–10. Table 4 of the *Candrārkī.*
Ff. 10–10v. Table 3 of the *Candrārkī.*
Colophon on f. 10v: iti caṃdrārkīkoṣṭaka saṃpūrṇam.

131. Khasmohor 5015(c). Ff. 1–2. 12 × 28 cm. 11 lines. Instructions in 26 verses.
Colophon on f. 2: iti caṃdrārkīsūtraṃ saṃpūrṇam.
After this is written:

daśārdrādyā⟨ḥ⟩ striyas tārā | viśākhādyā napuṃsakayoḥ |
tisras tataś ca mūlādyāḥ puruṣāḥ syuś caturddaśa || 1 |⟨|⟩
strīpuṃsayor mahāvṛṣṭiḥ || strīnapuṃsakayoḥ kvacit |
strīstriyoḥ śītalacchāyā yoge puruṣayor na ca || 2 ||
udayāstamane śukro || budhaś ca vṛṣṭikārakaḥ |
jalarāśisthitaś caṃdraḥ | pakṣāṃ[f. 2v]te saṃkrame[s] tathā | 3 |
budhaśukrasamīpasthaḥ karoty ekārṇavāṃ mahīṃ |
tayor aṃtargato bhānuḥ | samudram api śoṣayet || 4 ||
calaty aṃgārake vṛṣṭis tridhā vṛṣṭiḥ śanaiścare |
vārīpūrṇā⟨ṃ⟩ mahīṃ kṛtvā | paścāt saṃcarate maruḥ || 5 |⟨|⟩
atīcāragate jīve vakrībhūte śanimaṃgale |
hāhābhūtaṃ jagat sarva⟨ṃ⟩ rūṃḍamuṃḍā ca medinī || 6 ||
bhaumavakreṇāvṛṣṭiḥ || budhavakre janakṣayam ||
guruvakre jana[ḥ]pīḍā | śukravakre janaḥ sukhī || 7 ||
śanivakre bhaved rogaṃ | durbhikṣaṃ śatruvigrahaṃ || 8 || iti ||
śuddhito dyugaṇo hīnaḥ sa caivaṃ śuddhitas tyajet ||
pātya bāṇāṃgavahnibhyaḥ || ābdhikaḥ syād[d] ahargaṇaḥ || 1 ||
gatābdavṛṃdaṃ gaṇayec ca rudrair
adhaḥsthitaṃ vyomagirīṃdubhakta⟨ṃ⟩ 170 ⟨|⟩
yuktas tato cārugate dinasya || kharāmabhaktā tithir atra śeṣā |⟨| 1 || iti ||⟩

132. Khasmohor 5081. Ff. 1–10. 11 × 25.5 cm. Tables.
Ff. 1–4. Table 1 of the *Candrārkī*.
Ff. 4v–7. Table 2 of the *Candrārkī*.
Ff. 7–7v. Table 3 of the *Candrārkī*.
Ff. 7v–10. Table 4 of the *Candrārkī*.

133. Khasmohor 5082(e). 1 f. 11 × 25.5 cm. Tables. Part of manuscript 130?
Recto. Table 5 of the *Candrārkī*.
Verso is blank.

134. Khasmohor 5174(p). Ff. 5–7 and ⟨8⟩. 11 × 26.5 cm. Tables. Incomplete.
F. 5. End of Table 1 of the *Candrārkī* for 350 to 365 days.
Ff. 5–⟨8⟩. Table 2 of the *Candrārkī*.
F. ⟨8⟩. Table 4 of the *Candrārkī* for 1 to 27 days.
F. ⟨8v⟩ is blank.

135. Khasmohor 5174(t). F. 1. 11 × 27.5 cm. Tables. Incomplete.
F. 1. Table 3 of the *Candrārkī*.
F. 1v. Table 5 of the *Candrārkī*.

136. Khasmohor 5175(a). Ff. 1–3. 11.5 × 29 cm. Tables. Incomplete. Written to be hinged at the left.
F. 1: maṃdaphalaṃ traikyaṃ rāmavijasaṃskṛtā spaṣṭā.
Ff. 1–3v. Table 1 of the *Candrārkī* for 0 to 333 days.

137. Khasmohor 5254(a). Ff. 1–4 and 1 blank f. 11.5 × 26.5 cm. Tables. Incomplete.
Ff. 1–4. Table 4 of the *Candrārkī*.

138. Khasmohor 5254(e). Ff. 1–2. 11.5 × 26.5 cm. Tables. Incomplete.
F. 1. Table 5 of the *Candrārkī* for Śaka 1500 to 1680 (= A.D. 1578 to 1758) in steps of 20 years, and for 1 to 20 years.
Ff. 1v–2v contain mean motion tables (which are entries from the labdha tables of the *Mahādevī*) for the same units of time.
F. 1v Mars and Merçury.
F. 2. Jupiter and Venus.
F. 2v. Saturn and the lunar node.

139. Khasmohor 5254(i). Ff. 1–7. 11.5 × 26.5 cm. Tables. Incomplete.
Ff. 1–4. Table 1 of the *Candrārkī*.
Ff. 4v–7v. Table 2 of the *Candrārkī*.

A set of labdha tables for the *Mahādevī* for Śaka 1500 to 1700 (= A.D. 1578 to 1778) in steps of 20 years and for 1 to 20 years for Mars, Mercury, Jupiter, Venus, and Saturn; cf. manuscript 138, ff. 1v–2v.

Manuscript:

140. Khasmohor 5176(a). F. 1. 11 × 25.5 cm. Tables.
F. 1v. bottom: saṃvat 1635 śāke 1500 thakī mahādevīnā upakaraṇa sapta grahāṇāṃ 7.

The **ŚĪGHRASIDDHI**, **TITHISĀRIṆĪ**, or **CAMATKĀRACINTĀMAṆI** composed by Dinakara in 1583. *CESS* A3, 104b, and A5, 139b; and *SATE* 112–114. The first verse is:

> natvā gaṇeśaṃ gaganecarāṇāṃ
> gaṇaṃ pṛthag bhaktisamānato 'ham ||
> kurve camatkārakasārarūpaṃ
> sāraṃ gṛhītvā khalu śīghrasiddhim ||

Manuscript:

141. Khasmohor 5051. Ff. 1–18. 11.5 × 25.5 cm. Instructions in 21 verses on ff. 1–1v. Tables on ff. 1v–18.
Colophon on f. 1v: iti dinakaraviraciyāṃ tithisāriṇyāṃ camatkāraciṃtāmaṇyāṃ paṃcāṃgasiddhiḥ saṃpūrṇā.
F. 1v. Tables of the guṇakāras, kṣepakas, and rāmabījas for the śuddhi, the abdapa, and the manda according to the Brahmapakṣa.

Function	Guṇakāra	Kṣepaka	Rāmabīja
śuddhi	11;3,53,22,40	18;39,1,19,0	0;1,15,0,0
abdapa	1;15,31,17,17	0;42,35,23,0	0;0,10,0,0
manda	7;40,30,46,22	0;14,50,44,0	0;3,45,0,0

Table of kendras for tithis (27;59,33), nakṣatras (27;13,48), and yogas (29;16,1).
Table of guṇakas and ⟨kṣepakas⟩ of the śuddhi, abdapa, and manda according to the Digpakṣa (read Dṛkpakṣa).

Function	Guṇaka	⟨Kṣepaka⟩
śuddhi	11;3,53,3,28	18;34,9,3
abdapa	1;15,31,15,45	0;40,35,38
manda	7;40,30,38(?),31	0;6,25,33

Ff. 2–3v. Table 1 of the *Śīghrasiddhi* for 0 to 371.
Ff. 4–5v. Table 2 of the *Śīghrasiddhi* for 0 to 361.
Ff. 6–7v. Table 3 of the *Śīghrasiddhi* for 0 to 388.
Ff. 8–10v. Table 6 of the *Śīghrasiddhi.*
Ff. 11–13v. Table 7 of the *Śīghrasiddhi.*
Ff. 14–16v. Table 8 of the *Śīghrasiddhi.*
Ff. 17–17v. Table of the abdapa, tithidhruva, nakṣatrayogadhruva, manda, tithikendra, nakṣatrakendra, yogakendra, tithibhoga, nakṣatrabhoga, and yogabhoga for every year from Śaka 1556 to 1591 (= A.D. 1634 to 1669).
Ff. 17v. Tables 4 and 5 of the *Śīghrasiddhi.*
Table of bījas and of parameters according to the Āryapakṣa.
F. 18. Table of the gaṇakas of the tithidhruvas and tithikendras for 0 to 370 in steps of 10 as well as for 371, together with their cālakas, vārādis, and ghaṭyādis.
Table of the gaṇakas of the nakṣatradhruvas and nakṣatrakendras for 0 to 360 in steps of 9 as well as for 361, together with their various cālakas.
F. 18v. Table of the gaṇakas of the yogadhruvas and yogakendras for 0 to 387 in steps of 9 as well as for 388, together with their various cālakas.

The **RĀMAVINODA** in its koṣṭhaka form, composed by Rāma at Kāśī in 1590. *CESS* A5, 427a–428b; and *SATE* 114–118. The first verse is:

> gaṇapatiṃ abhivandya śrīśapādāravinda-
> smaraṇaviditatattvo 'nantajo rāmacandraḥ ||
> akabaranṛpaśākād vakti pañcāṅgapattraṃ
> grahagatisamavetaṃ rāmabhūpālatuṣṭya ||

Manuscript:

142. Puṇḍarīka jyotiṣa 49. 1 blank f,; and ff. 2–4 and ⟨5⟩. 10.5 × 22 cm. 8 lines. Instructions only. Incomplete.
F. 2 begins with || 4 ||, followed by 5a:

> saptāṃgarāmā nagabhūmayo bdhi.

Colophon on ff. 4v–⟨5⟩: iti śrīsamastasāmaṃtasīmaṃtinīsiṃd ūrapūradūrotsāraṇakāraṇaniṣkṛpak ṛpāṇadhārāvaluptamedinīpatimuṃḍama ṃḍalīmaṃḍitaṃ caṃḍīśamūrttimahīmaṃḍanaśrīm adakavaraśāhamahāmātyadhuryamahā rājādhirājaśrīrāmadāsadevakā[f. 5]ridevajñānaṃtabhaṭṭātmajarāmaviracite rāmavinodākhye paṃcāṃgagrahānayanādhikāraḥ.
F. ⟨5v⟩ is blank.

The tithi, nakṣatra, and yoga tables compiled by Harinātha at Kāśī on the basis of Rāma's *Rāmavinoda.* *CESS* A6. The first verse is:

dṛṣṭvā vinodaṃ rāmasya vinodaḥ kriyate 'dhunā ||
pañcāṅgaṃ harināthena śiṣyasantoṣahetave ||

Manuscript:

143. Khasmohor 5588. α ff. 1, ⟨2⟩, and 3–16; and β ff. 1–19 and 1f. 11.5 × 25 cm. Tables.

α. F. 1v. Table of tithis, vāras, and tithikendras for Śaka 1512 to 1720 (= A.D. 1590 to 1798) in periods of 16 years.

F. ⟨2⟩. Table of the same functions for 1 to 16 years.

F. ⟨2v⟩. Table of nakṣatras, vāras, and nakṣatrakendras for Śaka 1512 to 1800 (= A.D. 1590 to 1878) in periods of 24 years.

F. 3. Table of the same functions for 1 to 24 years.

F. 3v. Table of yogas, vāras, and yogakendras for Śaka 1512 to 1800 (= A.D. 1590 to 1878) in periods of 24 years.

F. 4. Table of the same functions for 1 to 24 years.

F. 4v. Table of the abdapas for Śaka 1512 to 1932 (= A.D. 1590 to 2010) in periods of 60 years.

F. 4v–5. Table of the abdapas for 1 to 60 years; the parameter for 1 year is 1;15,31,31,24.

F. 5v–8v. Table of the vāras and tithikendras for 0 to 407 tithis.

F. 8v–12. Table of the vāras and nakṣatrakendras for 0 to 408 nakṣatras.

F. 12v has just the skeleton of a table.

F. 13–16v. Table of the vāras and yogakendras for 0 to 406 yogas.

β. Ff. 1–7. Table of tithi equations for 0 to 29 horizontal, 0 to 59 vertical.

Ff. 7–13. Table of nakṣatra equations for 0 to 29 horizontal, 0 to 59 vertical.

Ff. 13–19. Table of yoga equations for 0 to 29 horizontal, 0 to 59 vertical.

F. 19v is blank.

The extra folium has a similar table for 0 to 29 horizontal, 19 to 23 vertical, on the recto; the verso is incomplete.

Price on α f.1: kī ||.

The **GRAHACĀRASAMUCCAYA**,an anonymous work on astrology and astronomy compiled in 1596. We describe only the astronomical parts.

Manuscript:

144. Khasmohor 4510. Ff. 1–5, ⟨6–38⟩, and additional folia. 29 × 13 cm. 27/28 lines. Writing parallel to the shorter edge. With marginalia. Bound in cardboard with a cloth cover bearing a flower design in pink and black and with a Persian-style flap. Ff. 1–⟨36v⟩ copied by Devarṣi Jyotiṣī for Lakṣmīcandra at Sāgānayari in Saṃ 1653, Śaka 1518 = A.D. 1596/7.

F. 3. An ayanāṃśakavidhi in 9 verses of which the first is:

grahamunividhuvahni3179miśritaḥ śākakālo
bhavati ayanavallī dvādaśādyaṃ ca guṇyaṃ ||
khakhasaramuni7500bhāgaṃ labhyate rāśim ādya
bhavati ayanamadhyaṃ krāṃtiṣaṃḍī viciṃtya || 1 ||

Ff. 3–3v. A saṅkrāntimahānakṣatradhruvāṃkakarttavyatā in 5 verses.
Ff. 3v–4. A ravibhagaṇavallī.
F. 4. A ravibījaśuddhamadhyama.
A candrabhagaṇavallī.
A caṃdrasabījamadhyama.
Ff. 4–4v. A kendrabhagaṇavallī.
Ff. 4v–5. A spaṣṭādhikāra following Āryabhaṭa.
F. 5. A tithyānayana.
A nakṣatrānayana.
Ff. 5–5v. A yogānayana.
Colophon on f. 5v: iti paṃcāṃgatithyādhikāra samāptaḥ.
F. 5v. Table of the mean motion of the Sun for 1 to 16 and 29 to 30 days.
Table of the mean motion of the Moon for the same days.
F. ⟨6⟩. Table of the kendra for the same days.
Table of Rāhu for the same days.

Ff. ⟨6–7⟩. Table of the equations of the Sun and their differences for 0° to 90° (the maximum is 2;10,49° at 90°).

Ff. ⟨7–7v⟩. Table of the equations of the Moon and their differences for 0° to 90° (the maximum is 5;1,42° at 90°).

(Ff. ⟨7v–9⟩. Texts on Mars, Mercury, Jupiter, Venus, Saturn, and Rānu. The first verse is:

bhūmiputro vijānīyā(d) rūponā bhagaṇair yutā
vedaghnā somaputre ca bhagaṇonā dvisaṃyutā || 1 ||

Colophon on f. ⟨9⟩: iti śrībrahmaguptācāryaviracite graṃthe khaṃḍakhādyake madhyamādhikāraḥ prathamaḥ. In fact, none of these verses seems to come from the *Khaṇḍakhādyaka*.

F. ⟨9v⟩. A bhaumādikaspaṣṭādhikāra.. It begins: śīghraṃ sthāpya maṃdaṃ pātya śīghrān madhye grahavisuddha ṣaḍūnaḥ.

Table of the mean motion of Mars for 1 to 16, 29 to 30, 353 to 355, and 383–385 days.

Ff. ⟨10–11v⟩. Table of the śīghra equations of Mars and their differences for 1° to 180° (the maximum is 40;37,44° at 131°).

Ff. ⟨11v–12v⟩. Table of the manda equations of Mars and their differences for 1° to 90° (the maximum is 11;10,0° at 90°).

Ff. ⟨12v–14v⟩. Table of the utkramajyās of Mars and their differences for 1° to 180° (the maximum is 40;30,33° at 44°). The utkramajyās are essentially the entries for the śīghra equations from 180° in reverse to 1°.

F. ⟨14v⟩. Table of the mean motion of Mercury for the same days.

Ff. ⟨14v–16v⟩. Table of the śīghra equations of Mercury and their differences for 1° to 180° (the maximum is 22;53,22° at 116°).

Ff. ⟨16v–17v⟩. Table of the manda equations of Mercury and their differences for 1° to 90° (the maximum is 4;27,52° at 90°).

Ff. ⟨18–20⟩. Table of the utkramajyās of Mercury and their differences for 1° to 180° (the maximum is 22;50,36° at 60°).

F. ⟨20⟩. Table of the mean motion of Jupiter for 1 to 16, 29 to 30, and 353–355 days.

Ff. ⟨20–22⟩. Table of the śīghra equations of Jupiter and their differences for 1° to 180° (the maximum is 11;35,37° at 101°).

Ff. ⟨22–23⟩. Table of the manda equations of Jupiter and their differences for 1° to 90° (the maximum is 5;6,16° at 90°).

Ff. ⟨23–25⟩Table of the utkramajyās of Jupiter and their differences for 1° to 180° (the maximum is 11;30,0° at 72°).

F. ⟨25⟩. Table of the mean motion of Venus for 1 to 16 and 29–30 days.

Ff. ⟨25–27⟩. Table of the śīghra equations of Venus and their differences for 1° to 180° (the maximum is 46;16,48° at 139°).

Ff. ⟨27–28v⟩. Table of the manda equations of Venus and their differences for 1° to 90° (the maximum is 2;10,49° at 90°).

Ff. ⟨28v–30v⟩. Table of the utkramajyās of Venus and their differences for 1° to 180° (the maximum is 46;11,30° at 38°).

Ff. ⟨30v–31⟩. Table of the mean motion of Saturn for 1 to 16 and 29 to 30 days.

Ff. ⟨31–33⟩. Table of the śīghra equations of Saturn and their differences for 1° to 180° (the maximum is 6;20,0° at 96°).

Ff. ⟨33–34⟩. Table of the manda equations of Saturn and their differences for 1° to 90° (the maximum is 7;39,25° at 90°).

Ff. ⟨34–36v⟩. Table of the utkramajyās of Saturn and their differences for 1° to 180° (the maximum is 6;20,23° at 81°).

Colophon on f. ⟨36v⟩: iti śrīṣaṃḍaṣādyake śani utkram⟨ā⟩ptajyā saṃpūrṇaṃ koṭhā 2664 paryaṃta.

Post-colophon: saṃvat 1653 varṣe śāke 1518 pravarttamāne sāgānayarimadhye liṣitaṃ jyotiṣī devarṣi ācāryaśrī5lakṣmīcandrārthaṃ.

Ff. ⟨36v–37v⟩. Various notes in bhāsā and in Saṃskṛta including tables of the week-days on which the Sun enters each of the 12 zodiacal signs and of the 28 nakṣatras and texts on karaṇas and retrogressions.

F. ⟨38⟩. Verses on geography.

graharaṃdhracāraṇapāvaka 3299-
bhūgolārdhaṃ vadaṃti gaṇ⟨i⟩tajñāḥ |
rasarajanīkara16bhaktā [||]
avaṃtīlaṃkāṃtaraṃ mānaṃ || 1 ⟨||⟩
ujjeṇīruhitakakukhya-
jamunāhimagirinivāsīnāṃ
deśāṃtaraṃ [||] na kāryā
talleṣā madhyasaṃsthitā || 2 ⟨||⟩

ujjenījāmyottarareṣāyāṃ prāk ṛṇaṃ dhanaṃ paścāt ||
deśāṃtarabhuktivadhāt bhūparināho dhṛtaḥ kalādyāptaṃ || 3 ⟨||⟩

pāṭhāṃtaraṃ||

purī rākṣasāṃ devakanyātha kāṃtī
sitaḥ parvvataḥ parjvalī vatsagulmāṃ ||
purī cojjayanyāhvayā gargarāṭaṃ
kurukṣetramerur bhaven madhyareṣā|| 4 ||
avaṃtiyāmyottarayojanāni
saṃguṇya vāṇai5 rasa6bhājitāni ||
hīnādhikāny aṃganakheṣu 206 kuryāc
caṃdrābdhi41bhaktā viṣ⟨u⟩vatprabhā⟨ḥ⟩ syuḥ || 5 ||
ayanalavadinaiḥ prāṅ meṣasaṃkrāṃtikālād
bhavati divasamadhye yā prabhākṣ[y]aprabhā [||] sā ||
daśa10gaja8daśa10nighnī sākṣabhāṃtyā tribhaktā
pratigrahacarakhaṃḍāny āyanāṃśāḍhyabhānoḥ || 6 ⟨||⟩
sarāṃśa15sahitā rūpa[ṃ]yugmāgnau 321 jojanāni ca |
purutaḥ yoginīcakraṃ aṃtaraṃ rākṣasā⟨ṃ⟩ purī || 7 ||
naṃdākṣā 59 bhujagā 8 raveḥ śaśigatiḥ khāṃkādrayo 790 ⟨'⟩kṣāgnayaḥ 35
tuṃga[ḥ]syāṃgakalā 6 kuveda41vikalāḥ pātasya rāmā 3 bhavā 11 ⟨||⟩
māheyasya mahīguṇā 31 rasakarā 26 jñasyeṣusiddhā 245 radā⟨ḥ⟩ 32
paṃce5jyasya sitasya ṣaḍṇava96mitā aṣṭau 8 śaner dve 2 kale | 1 ⟨|⟩
vrahmatulyenoktaṃ ||

Verse 4 is I 14, verse 6 is II 19, and the last verse is I 13 of Bhāskara's *Karaṇakutūhala*.

An anonymous set of tables following the Saurapakṣa; their epochs are 1598 or 1618.

Manuscript:

145. Khasmohor 5349(a). Ff. 1–6. 11 × 26 cm. Tables.

F. 1v. Table of the śuddhi, the abdapa, the Moon, and its kendra from Śaka 1520 to 1700 (= A.D. 1598 to 1778) at intervals of 20 years.

Table of the same functions for 1 to 20 years.

Ff. 2–2v. Table of the śuddhi, the adhimāsas (to Śaka 1669 = A.D. 1747), and the abdapa from Śaka 1582 to 1724 (= A.D. 1660 to 1802).

Ff. 3–3v. Table of the mean longitudes with their bījas of the Moon, its apogee, Mars, Mercury, Jupiter, Venus, Saturn, and Rāhu from Śaka 1540 to 1680 (= A.D. 1618 to 1758) in intervals of 20 years.

Table of the mean longitudes of the same entities for 1 to 20 years.

Ff. 4–4v. Table of the mean longitudes of the Sun, the Moon, its apogee, Mars, Mercury, Jupiter, Venus, Saturn and Rāhu for 1 to 27 avadhis.

F. 5. Table of the mean longitudes of the same entities for 1 to 13 days.

Table of the mean daily motions of the same entities (Rāhu is placed between the Moon's apogee and Mars).

Ff. 5v–6. Table of the mean motions of the same entities (with Rāhu at the end) for 1 to 60 ghaṭīs.

F. 6v. Table of the equations of the manda of the Sun, the Moon, Mars, Mercury, Jupiter, Venus, and Saturn for 10° to 90° of argument at intervals of 10°.

Table of the equations of the śīghra of Mars, Mercury, Jupiter, Venus, and Saturn for 10° to 180° of argument at intervals of 10°.

The **TITHIPATRACINTĀMAṆI** whose manuscript was copied in Saurāṣṭra in 1604.

Manuscript:

146. Khasmohor 5294. Ff. 2–13. 12 × 22 cm. Incomplete. Copied by Govarddhana, the son of Śrīnātha(?) of the Nāgarajñāti, a resident of Jīrṇadurga, for Dāmodara Miśra in 1604/5.

F. 2 begins: na(?)krāgnikoṣṭe pūrvalikhitāni

> nagāgnisaptabhi737r bhaktaṃ pṛthag maṃdaṃ phalonitaṃ |
> dhruvanāḍyādinā yuktaṃ nije keṃdraṃ sphuṭaṃ bhavet || 8 ||

Colophon on f. 2v: iti tithipatraciṃtāmaṇiḥ samāptaḥ.

Post-colophon: svasti saṃvat 1661 varṣe śrīmajjīrṇadurganivāsinā nāgarajñātinā vyava(?)śrīhāthī(?)sutena govarddhanena tithipatraciṃtāmaṇir alekhi samāptaḥ || || || miśraśrīdāmodarapaṭhanārthaṃ.

Ff. 3–6. Table of 0 to 371 tithis indicating their kendras and vārādis.

Ff. 6–9. Table of 0 to 361 nakṣatras indicating their kendras and vārādis.

Ff. 9v–13. Table of 0 to 388 yogas indicating their kendras and vārādis.

Ff. 13. Table of the weekdays on which the Sun enters each zodiacal sign.

Table of the weekdays on which the Sun enters each of the nakṣatras.

Table of abdapas according to the Brāhmapakṣa.

Colophon on f. 13: iti tithipatraciṃtāmaṇiḥ samāptaḥ.

F. 13v. Table of kendraphalas and hāras for 0 to 27 tithis and nakṣatras and 0 to 29 yogas.

Price on f. 2v: kī ·=.

The **KHAGATARAṄGIṆĪ** composed by Goparāja according to the Saurapakṣa; its epoch is 1608. *CESS* A6. The first verse is:

praṇipatya gaṇādhīśaṃ goparājābhidhaḥ sudhīḥ ||
saurānusāraṇīṃ spaṣṭāṃ kurve khagataraṅgiṇīṃ ||

Manuscript:

147. Khasmohor 5049(b). Ff. 55b–64 and 67–93. 11 × 25 cm. 17 lines and tables. Incomplete.

Colophon on f. 56: iti spaṣṭagatiprakāraḥ.

F. 56v. Table of abdapeṣu saṃskārapalāni (correction in palas to the lengths of days). The argument is 0 to 12 (zodiacal signs) vertical and 0 to 27 (parts) horizontal. There are maxima at parts 16 to 18 of zodiacal sign 2 (Taurus)(2;1 ghaṭikās), and at parts 5 to 7 of sign 10 (Aquarius) (2;13 ghaṭikās), minima at part 20 of sign 6 (Libra)(0 ghaṭikās) and between part 18 of sign 12 (Aries?)(-0;2 ghaṭikās) and part 19 (+0;1 ghaṭikās).

Ff. 57–57v. Table of mean motions of the śuddhi, abdapa, Moon, Mars, Mercury, Jupiter, Venus, Saturn, the Moon's apogee, and its node; given are the kṣepakas and the mean motions for 1 to 9 years, for 10 to 90 years in steps of 10, and for 100 to 00 years in steps of 100.

Ff. 58–58v. Table of the mean motions of the same entities for 1 to 9 days, for 10 to 90 days in steps of 10, and for 100 to 300 days in steps of 100.

F. 59. Manda equation and velocity equation of the Sun for 0° to 90°. The solar apogee is at Gemini 17;16,48°, the maximum equation 2;10,34° at 90°.

F. 59v. Table of the manda equation and velocity equation of the Moon for 0° to 90°. Maximum equation 5;2,48° at 90°.

Ff. 60–60v. Table of the śīghra equation, the manda equation, and the hypotenuse of Mars for 0° to 180°. Maximum śīghra equation 40;16 at 130° to 132° and maximum manda equation 11;32° at 93° to 98°.

Ff. 61–61v. Similar table for Mercury. Maximum śīghra equation 21;31° at 108° to 113° and maximum manda equation 4;28° at 92° to 93°.

Ff. 62–62v. Similar table for Jupiter. Maximum śīghra equation 11;31° at 100° to 102° and maximum manda equation 5;6° at 91° to 95°.

Ff. 63–63v. Similar table for Venus. Maximum śīghra equation 46;24° at 136° to 137° and maximum manda equation 1;45° at 85° to 97°.

Ff. 64–64v. Similar table for Saturn. Maximum śīghra equation 6;22° at 94° to 98° and maximum manda equation 7;40° at 92° to 96°.

F. 67. Sāyana dinamānapatra for Burahānapura, where the akṣabhā is 4;30 digits. The argument is 0 to 11 zodiacal signs vertical and 0 to 29 degrees horizontal. The maximum daylight is 33;12 ghaṭikās at Cancer 0°, the minimum 26;48 ghaṭikās at Capricorn 0°.

F. 67v. Similar table for places where the akṣaprabhā is 5;0. The maximum is 33;33 ghaṭikās at Cancer 0°, the minumum 26;27 ghaṭikās at Capricorn 0°.

F. 68 Similar table for Citrakūṭa, Udayapura, and Sīrohī where the akṣaprabhā is 5;30. The maximum is 33;54 ghaṭikās at Cancer 0°, the minumum 26;6 ghaṭikās at Capricorn 0°.

F. 68v. Similar table, but giving half-daylights, for places where the akṣabhā is 6;30. The maximum is 17;19 ghaṭikās at Cancer 0°, the minumum 12;41 ghaṭikās at Capricorn 0°.

F. 69. Similar table giving daylights for Kurukṣetra where the akṣabhā is 7;0. The maximum is 34;58 ghaṭikās at Cancer 0°, the minumum 25;2 ghaṭikās at Capricorn 0°.

F. 69v. Similar table for Lāhora where the akṣabhā is 7;30. The maximum is 35;20 ghaṭikās at Cancer 0°, the minumum 24;40 ghaṭikās at Capricorn 0°.

F. 70. Similar table for Kāśmīra where the akṣabhā is 8;0. The maximum is 35;41 ghaṭikās at Cancer 0°, the minumum 24;19 ghaṭikās at Capricorn 0°.

F. 70v. Similar table for Kābilanagara where the akṣabhā is 8;30.

F. 71–71v. Verses in bhāṣā and Saṃskṛta on computing the ascendent. The first verse is:

tātkāliko 'rko 'yanabhāgayuktas
tadbhogyabhāgair udayo hataś ca ⟨||⟩
khāgnyu30ddhṛtas taṃ ravibhogyakālaṃ
viśodhayed iṣṭaghaṭīpalebhyaḥ ⟨||⟩ 1 ⟨||⟩

Colophon on f. 71v: iti lagnaspaṣṭaḥ.

F. 72. Table of the nirayanalagnasāraṇī for a place where the akṣabhā is 4;30 digits. The argument is 0 to 11 (zodiacal signs) vertical and 0 to 29 (degrees) horizontal. In this and the following tables 0 (i.e., the vernal equinox) occurs at sign 11 (Pisces) 10°, so that the precesseion is 20°.

F. 72v. Similar table for a place where the akṣabhā is 5;0.

F. 73. Similar table for a place where the akṣabhā is 5;30.

F. 73v. Similar table for a place where the akṣabhā is 6;0.

F. 74. Similar table for a place where the akṣabhā is 6;30.

F. 74v. Similar table for a place where the akṣabhā is 7;0.

F. 75. Similar table for a place where the akṣabhā is 7;30.

F. 75v. Similar table for a place where the akṣabhā is 8;0.

F. 76. Similar table for a place where the akṣabhā is 8;30.

Colophon at bottom of f. 76: iti lagnasāraṇī samāptā.

F. 76v. Table for computing the cusps of the twelve bhāvas from the ascendent. Argument 0 to 11 (zodiacal signs) vertical and 0 to 29 (degrees) horizontal.

F. 77. Several short pieces in bhāṣā and Saṃskṛta on computing the cusps of the bhāvas. It begins; lagnanī rāśi smśa naiṃ ko.

F. 77v. Table of the nirayanadaśamabhāva in units of 6°. Argument 0 to 11 (zodiacal signs) vertical and 0 to 29 (degrees) horizontal. The entry is 0;0 at sign 8 (Sagittarius) 11°.

Ff. 78–78v. Continuation of short pieces on f. 77.

Ff. 79–79v. Table for ṣaḍaṃśabhāva. The argument is as in the immediately preceding tables. The entry at 0^s 0° is 15;0,0 (× 6 = 90), at 2^s 0° 12;44,32 (× 6 = 76;15,12), and at 10^s 0° 17;15,27 (× 6 = 103;32,42).

F. 80. Text on bhāvas.

F. 80v. Two tables entitled: viśvānayanārthaṃ grahasaṃdhyaṃtarakoṣṭakāḥ. The argument in the first is 1 to 15 horizontal and 12;0 to 14;45 vertical in steps of 0;15; the argument in the second is 1 to 17 horizontal and 15;0 to 17;45 vertical in steps of 0;15.

Colophon on f. 81: iti bhāvāḥ saṃpūrṇāḥ

F. 81. Text on computing aspects of the planets. The first verse is:

> uktāni yasmād bahudhā phalāni
> vyomaukasāṃ dṛṣṭisamudbhavāni ⟨||⟩
> tasmāt pravacmy ānayanaṃ ⟨ca⟩ dṛṣṭer
> horāvidāṃ dṛkphalanirṇayāya ⟨||⟩ 1 ⟨||⟩

F. 81v. Table of an unspecified planet's (the Moon's?) aspects. The argument is 1 to 9 vertical and 0 to 29 horizontal.

F. 82. Similar table of Mars' aspects.

F. 82v. Similar table of Jupiter's aspects.

f. 83. Similar table of Saturn's aspects.

Colophon on f. 83v: iti grahāṇāṃ dṛṣṭiḥ.

Ff. 83v–87. Short astrological pieces.

Ff. 87v–93v. Verses on dvādaśabhāvavicāra. A second scribe has written at the top of f. 87v: śrīpatipaddhati.

The first verse is:

> udayaṃ janmabhaṃ lagnaṃ horā śīrṣaṃ dyutas tanuḥ ⟨||⟩
> śarīraṃ pūrvataḥ kalpaṃ svasthānaṃ mūrttibhājanaṃ ⟨||⟩ 1 ⟨||⟩

It ends on f. 93v in verse 55:

> ṣaṣṭāṣṭame yadā caṃdra ravir bhavati saptame ⟨||⟩
> pitṛmātṛdhanaṃ

Neither verse is found in the *Śrīpatipaddhati.*

A derivative of the tables composed by Harideva in 1610 on the basis of the Babylonian Goal-year Periods.

Manuscript:

148. Khasmohor 5059(b). F. 2 (renumbered 1). 11 × 26 cm. Tables. Incomplete.

F. 2. Blank.

F. 2v. Table of Mercury for the years (Śaka) 1532 to 1545 (= A.D. 1610 to 1623). Each year has 3 columns corresponding to the 3 synodic periods of Mercury in each year (sometimes a fourth column is also used); each line corresponds to a Greek-letter phenomenon, in order: first station (Ψ), second station (Φ), last visibility in the West (Ω), first visibility in the East (Γ), last visibility in the East (Σ), and first visibility in the West (Ξ). Each set of columns is valid not only for the first year, but for 46 years later (Śaka 1578 to 1591) and 46 years after that (Śaka 1624 to 1637). The entries are in degrees and minutes.

The **GAṆITAMAKARANDA** composed by Rāmadāsa Dave at Vaḍanagara with 1618 as epoch. *CESS* A5, 490a–490b. The first verse is:

gaṇeśaṃ giraṃ kālam īśaṃ natvā
yataḥ prasphuṭaṃ kheṭakarma pravakṣye ||
avantyuttare saṃsthitā śuddhadantī
maror maṇḍale ṣaṭtribhir yojanaiś ca ||

Manuscript:

149. Khasmohor 5059(a). Ff. 1–9. 11 × 26 cm. 16–18 lines or tables. With some marginalia. Manuscript 148 is between ff. 1 and 2. Copied by Ṛṣi Thaṣa(?), the pupil of Ṛṣi Kaṭājī, the pupil of Ṛṣi Gamvajī(?), at Medanīpurī on *ca.* 7 December 1656.

Colophon on line 7 of f. 1: iti śrīgaṇitamakarande rāmadāsakṛtai miśrākādhyāyaḥ || prathamaḥ || saṃpūrṇaḥ ||

F. 1v ends, in line 11, in 2, 4c: triṃsatkoṣṭatithau 30 bhayogapritibhaṃ 27 keṃdre svayuk sa.

F. 2. Table of the weekday on which the Sun enters each of the 12 zodiacal signs.

Table of the weekday on which the Sun enters each of the 27 nakṣatras.

Table of yearly parameters of the śuddhi, abdapa, tithidhruva, nakṣatrayogadhruva, tithikendra, nakṣatrakendra, yogakendra, tithivāra, nakṣatravāra, and yogavāra.

Table of epoch entries for these functions.

Table of the cakras of tithis, nakṣatras, and yogas.

Table of tithikendrakṣepakas for 1 to 13 months.

Table of tithikendrakṣepakas for 1 to 13 months.

Table of nakṣatrakendrakṣepakas for 1 to 14 months (the entry for 1 month is 0;0,0).

F. 2v. Table of nakṣatravārādikṣepas for 1 to 14 months (the entry for 1 month is 0;0,0).

Table of yogakendrakṣepakas for 1 to 15.

Table of yogavārādikṣepas for 1 to 15.

Ff. 2v–3v. Table of tithikendras for 0° to 55° vertical in steps of 5° and for 0 to 29 horizontal.

Ff. 3v–4v. Similar table of nakṣatrakendras.

Ff. 4v–5. Similar table of yogakendras, but the vertical argument is 0° to 35° in steps of 5°, and then 50°, 55°, 0°, and 5°. At the bottom of f. 5 is the colophon: iti śrīgaṇitamakaraṃde rāmadāsakṛtau tithyādipañcāṅgasiddhiḥ dvitīyo dhyāyaḥ || 2 || followed by the beginning of adhyāya 3.

F. 5v ends with the colophon: iti śrīgaṇitamakaraṃde rāmadāsakṛtau grahasiddhiḥ spaṣṭādhikārahs tṛtīyaḥ.

F. 6. Table of the mean yearly motions of the epact, the lord of the year, the Sun, the Moon, Mars, Mercury, Jupiter, Venus, Saturn, the lunar node, and the lunar apogee.

Table of the epoch positions of the same entities.

Table of the longitudes of the apogees of the Sun, Mars, Mercury, Jupiter, Venus, and Saturn.

Table of the Rāmabījas of the Sun, the Moon, Mars, Mercury, Jupiter, Venus, Saturn, the lunar node, and the lunar apogee.

Table of the mean daily motions of the same entities.

Table of the mean motions of the same entities in a period of 14 days.

Table of the manda equations of the Sun for arguments of 10,20,30,...90 degrees.

Table of the manda equations of the Moon for the same arguments.

Table of the manda equations of Mars for the same arguments.

Table of the manda equations of Mercury for the same arguments.

Table of the manda equations of Jupiter for the same arguments.

Table of the manda equations of Venus for the same arguments.

Table of the manda equations of Saturn for the same arguments.

Table of the śīghra equations of Mars for arguments of 10,20,30,...180 degrees.

Table of the śīghra equations of Mercury for the same arguments.

Ff. 6–6v. Table of the śīghra equations of Jupiter for the same arguments.

F. 6v. Table of the śīghra equations of Venus for the same arguments.

Table of the śīghra equations of Saturn for the same arguments.

Ff. 6v–7. Table of the elongations from the Sun at the occurrences of the Greek-letter phenomena of Mars, Mercury, Jupiter, Venus, and Saturn.

Colophon on f. 6v: iti śrīśrīgaṇitamakaraṃde samadākṛtau grahasiddhiḥ spaṣṭādhikārahs tṛtīyaḥ || 3 ||

F. 7 begins with 4, 1a: ṣaṣṭighnā 60.

Colophon on f. 9v: iti śrīgaṇitamakaraṃde rāmadāsakṛtau vakramārgrādi udayāstagrahayutivyatipātau caṃdradarśano dhyāyo ṣṭamaḥ.

Post-colophon: likhitaṃ ṛṣi śrīśrīśrīśrīśrīgamvājī(?)

tacchiṣya ṛṣi śrīśrīśrīketājī tacchiṣya ṛṣi thaṣena pravācyanārtham | śrīmedanīpuryyāṃ || saṃvat 1713 varṣe posasudi 2 di(ne).

The **KHEṬAKAUTŪHALA** composed by Sūrajit at a royal city (Ahmadabad?) on the bank of the Kāśyapanandinī (Sabarmatī) in 1619. The first verse is:

namaskṛtya durgāṃ śivaṃ vighnarājaṃ
guruṃ śrīdharaṃ khecarāṃś cārkapūrvān ||

pravakṣye sphuṭaṃ kheṭakautūhalākhyaṃ
vināyāsasiddhipradaṃ sūranāmā ||

Manuscript:

150. Khasmohor 5607. Ff. 1–4. 16 × 29 cm. 15 lines. Edges of leaves irregular and tattered. With some marginalia.

Colophon on f. 4: iti śrīdāmodarātmajasiṃhajitsutasūrajidviracite kheṭakautūhale asmād udvaritayoḥ avadhigraṃthayo grahasādhanādhyāyaḥ || samāpto kheṭakautūhalaḥ.

An udāharaṇa on Sūrajit's *Kheṭakautūhala* composed in 1628 or 1629. It begins: atha ṣeṭakutūhale udāharaṇam aharggaṇa 363.

Manuscript:

151. Khasmohor 5602. Ff. 1–3. 16 × 30 cm. 21–24 lines. With marginal notes. Incomplete.

F. 3v ends in the section on Venus: spaṣṭaḥ śukraḥ gatispaṣṭīkaraṇam.

The **GRAHALĀGHAVASĀRIṆĪ** composed by Nīlakaṇṭha in about 1630. *CESS* A3, 174b; A4, 142a; and A5, 184b–185a. The first verse is:

gaṇādhīśaṃ namaskṛtya grahalāghavasāriṇī |
kriyate nīlakaṇṭhena śiṣyasantoṣahetave ||

Manuscripts:

152. Puṇḍarīka jyotiṣa 36. Ff. 1–18. 11 × 26 cm. 12 lines; and tables. Copied by Gopīnātha for Gokulanātha. Formerly property of Gokulanātha, the son of Śambhūnātha Jośī.

Colophon on f. 3: iti śrīmatsakalagaṇakasārvabhaumaśrīmachivagaṇakasutena
nīlakaṃṭhena viracitāyāṃ grahalāghavasāraṇyāṃ bhaumādipaṃcagrahaspaṣṭīkaraṇaṃ saṃpūrṇaṃ.

f. 3v. Blank.

Ff. 4–4v. Tables of the mean motions of the Sun, the Moon, the Moon's apogee, Rāhu, Mars, Mercury's kendra, Jupiter, Venus' kendra, and Saturn for 1 to 9 days and for 10 to 60 days in steps of 10.

F. 4v. Table of the mean motions of the same nine entities and of the Moon's kendra, Mercury's śīghrocca, and Venus' śīghrocca for 7 days.

Table of the yearly motions with their bījas of the Lord of the year (1;15,31,31,24 days), the Epact (11;3,53,15 tithis), the Sun (0^s 0;0,0,0°), the Moon (4^s 12;46,41,0°), the Moon's kendra (3^s 2;5,40,42°), Rāhu (19;21,12,12°), Mars (6^s 11;24,9,0°),

Mercury (1^s 24;45,22,48°), Jupiter (1^s 0;21,3,26°), Venus (7^s 15;11,44,25°), and Saturn (12;12,53,36°).

Table of the mean motions of the Sun, the Moon, the Moon's apogee, Rāhu, Mars, Mercury, Jupiter, Venus, and Saturn in a quarter, a half, three quarters, and a complete day.

F. 5. Table of the mean motions of the Sun for 1 to 40 cycles (of 4016 days = 11 "years) and for 1 to 9 days, 10 to 90 days in steps of 10, 100 to 900 days in steps of 100, and 1000 to 4000 days in steps of 1000.

F. 5v. Table of the mean motions of the Moon set up like that for the Sun.

F. 6. Table of the mean motions of the Moon's apogee set up in the same way except only for aṅkas 1 to 39.

F. 6v. Table of the mean motions of Rāhu set up in the same way as that for the Moon's apogee.

F. 7. Table of the mean motions of Mars set up as is the preceding.

F. 7v. Table of the mean motions of Mercury's kendra set up as is the preceding.

F. 8. Table of the mean motions of Jupiter set up as is the preceding.

F. 8v. Table of the mean motions of Venus' kendra set up as is the preceding.

F. 9. Table of the mean motions of Saturn set up as is the preceding.

F. 9v. Table of the manda equation and correction to the velocity of the Sun for 0° to 90° of argument; the maximum is 2;10,54° at 90°.

F. 10. Similar table for the Moon; the maximum is 5;1,40° at 90°.

F. 10v. Similar table for Mars; the maximum is 13;0,0° at 90°.

F. 11. Similar table for Mercury; the maximum is 3;36,0° at 90°.

F. 11v. Similar table for Jupiter; the maximum is 5;42,0° at 90°.

F. 12. Similar table for Venus; the maximum is 1;30,0° at 90°.

F. 12v. Similar table for Saturn; the maximum is 9;18,0° at 90°.

F. 13–13v. Table of the śīghra equation and correction to the velocity of Mars for 0° to 180° of argument; the maximum is 40;0,0° at 135°.

Ff. 14–14v. Similar table for Mercury; the maximum is 21;12,0° from 105° to 120°.

Ff. 15–15v. Similar table for Jupiter; the maximum is 10;48,0° at 105°.

Ff. 16–16v. Similar table for Venus; the maximum is 46;6,0° at 135°.

Ff. 17–17v. Similar table for Saturn; the maximum is 5;42,0° from 90° to 105°.

F. 18. Tables of the elongations from the Sun required for the occurrence of the Geek-letter phenomena for each of the planets.

Colophon on f. 18: iti śrīmatsakalagaṇakasārvabhaumaśrīmachivagaṇakasutena

viracitāyāṃ grahalāghavasāriṇyāṃ bhaumādipaṃcagrahaspaṣṭīkaraṇaṃ saṃpūrṇatām agamat.

Post-colophon: li. gopīnāthena gokalanāthasya paṭhan(ārthaṃ || vismaraṇārthaṃ ca).

F. 18v. Blank.

Ownership note on f. 1: idaṃ jośī śaṃbhūnāthasutagokulanāthasya pustakaṃ varttate.

153. Puṇḍarīka jyotiṣa 41. Ff. 1–14. 14 × 30 cm. Tables.

F. 1. Table of the mean motions of the Sun as on f. 5 of manuscript 152.

F. 1v. Table of the mean motions of the Moon as on f. 5v of manuscript 152.

F. 2. Table of the mean motions of the Moon's apogee as on f. 6 of manuscript 152.

F. 2v. Table of the mean motions of Rāhu as on f. 6v of manuscript 152.

F. 3. Table of the mean motions of Mars as on f. 7 of manuscript 152.

F. 3v. Table of the mean motions of Mercury's kendra as on f. 7v of manuscript 152.

F. 4. Table of the mean motions of Jupiter as on f. 8 of manuscript 152.

F. 4v. Table of the mean motions of Venus' kendra as on f. 8v of manuscript 152.

F. 5. Table of the mean motions of Saturn as on f. 9 of manuscript 152.

F. 5v. Table of the manda equation of the Sun as on f. 9v of manuscript 152.

F. 6. Table of the manda equation of the Moon as on f. 10 of manuscript 152.

Ff. 6v–7. Table of the manda equation of Mars as on f. 10v of manuscript 152.

Ff. 7–7v. Table of the manda equation of Mercury as on f. 11 of manuscript 152.

Ff. 7v–8v. Table of the manda equation of Jupiter as on f. 11v of Puṇḍarīka Jyotiṣa 36.

Ff. 8v–9. Table of the manda equation of Venus as on f. 12 of manuscript 152.

Ff. 9–9v. Table of the manda equation of Saturn as on f. 12v of manuscript 152.

Ff. 9v–11. Table of the śīghra equation of Mars as on ff. 13–13v of manuscript 152.

Ff. 11–11v. Table of the śīghra equation of Mercury as on ff. 14–14v of manuscript 152.

Ff. 11v–12v. Table of the śīghra equation of Jupiter as on ff. 15–15v of manuscript 152.

Ff. 12v–13v. Table of the śīghra equation of Venus as on ff. 16–16v of manuscript 152.

Ff. 13v–14. Table of the śīghra equation of Saturn as on ff. 17–17v of manuscript 152.

F. 14v. Table of the elongations from the Sun required for the occurrence of the Greek-letter phenomena as on f. 18 of manuscript 152.

Colophon on f. 14v: iti śrīmatsakalagaṇakasārvabhaumaśrīmacchivagaṇakasutena

nīlakaṃṭena viracitāyāṃ grahalāghavasāraṇyāṃ bhaumādipaṃcagrahaspaṣṭīkaraṇaṃ saṃpūrṇatām agamat.

After this is written:

sūryarkṣato starkṣam athodayarkṣaṃ
khamadhyagarkṣaṃ gaṇaye⟨t⟩ krameṇa ||
hīnaṃ mahī1tithya15hi8bhir niśārdha-
nighnaṃ nagāptaṃ niśi bhuktaghaṭyaḥ || 1 ||
sūryarkṣād astabhaṃ sūryabhān
maulibhaṃ gaṇyaṃ saptahīnaṃ ca śeṣakaṃ ||
dviguṇaṃ dvayahīnaṃ ca
gatā rātriḥ sphuṭā bhavet ||1 ||

The **JAGADBHŪṢAṆA** completed by Haridatta under Jagatsiṃha of Mewar in 1638. *CESS* A6; *SATIUS* 55b–59b. The opening verses are:

praṇamyādidevaṃ vibhuṃ nārasiṃhaṃ
grahān bhāskarādyān gaṇeśaṃ bhavānīm |
graharkṣādītithyādipañcāṅgasiddhyai
karomi prabandhaṃ jagadbhūṣaṇākhyam || 1 ||
devendras tridaśeṣu vṛkṣanicaye kalpadrumas tārakāsv
induḥ parvatasaṃcaye suragirī ratneṣu cintāmaṇiḥ |
gaṅgā satsaritāṃ gaṇe maragavī goṣu dvipānāṃ gaṇe
svarṇāgo 'khilarājavaṃśanikare śrīsūryavaṃśas tathā || 2 ||

Manuscripts:

154. Khasmohor 5083. Ff. 1–8. 11 × 27 cm. 14 lines. Verses only. Copied in 1741/2.

Colophon on ff. 8–8v: iti śrīgaṇakacakracūḍāmaṇibhaṭṭaharajītanujahara-dattabha[f. 8v]ṭṭaviracite niṣalarājamaṃḍalīmaulamaṃḍitapratāpanikaravidhva-staripumaṃlavivadhavidyāvinodaraśikamahārājādhirājamahārāṇāśrījagatsihanā-māṃkite
śrījagadbhūṣaṇe paṃcāṃgaspaṣṭīkaraṇādhikāra paṃcama 5.
After this is written:

graṃthe smīn yad vīlīkaṃ syāj jagadbhūṣaṇasaṃjñake
suddhībhiḥ sodhanīyaṃ tat suvicārya punaḥ punaḥ 1
pūṇatarkkaśarabhūmitaśake 1560 kārttike dhavalapakṣaśamete
paṃcamitithiyute budhavāre bhūṣaṇaṃ samakarod dharattadaḥ 2
iti śrīmajagadbhūṣaṇa saṃpūrṇaḥ

Post-colophon: 1798.

155. Khasmohor 5349(b). 69 ff. 11 × 26 cm. Tables. Copied by Ṛṣaladāsa or Ṛṣalalāla Yati, the pupil of Lakṣmīcandra Vācaka, at Savāī Jayapura between *ca.* 15 February 1757 and *ca.* 20 June 1757. Incomplete.

Ff. ⟨1–10⟩. Table 1 of the *Jagadbhūṣaṇa* (Mars) for Śaka 1560 to 1717 (= A.D. 1638 to 1795).

Colophon on f. ⟨10v⟩: iti baumaspaṣṭaḥ samāptaḥ.

Post-colophon: li° ṛṣaladāsa gaṇiḥ.

Ff. ⟨11–22⟩. Table 2 of the *Jagadbhūṣaṇa* (Mercury) for Śaka 1560 to 1697 (= A.D. 1638 to 1775).

Colophon on f. ⟨22⟩: iti śrījagadbhūṣaṇau budhaḥ sapūrṇaḥ.

Post-colophon: savaījayapurame li° ṛṣalalālena yatinā matinā.

Ff. ⟨22v–32v⟩. Table 3 of the *Jagadbhūṣaṇa* (Jupiter) for Śaka 1560 to 1725 (= A.D. 1638 to 1803).

Colophon on f. ⟨32v⟩: iti śrījagadbhūṣaṇaśāstre | guruḥ spaṣṭakoṣṭakāni saṃpūrṇāni jātāni.

Post-colophon: saṃvat 1813 varṣe kārttikamāse śuklapakṣe daśāmyāṃ ⟨as⟩au liṣitā ṛṣalalālena yatinā matinā | śrīsavāījayapuramadhye.

Ff. ⟨33–61⟩. Table 4 of the *Jagadbhūṣaṇa* (Venus) for Śaka 1560 to 1785 (= A.D. 1638 to 1863).

Colophon on f. ⟨61⟩: iti śrījagadbhūṣaṇaśāraṇyāyāṃ śukraspaṣṭaḥ samāptaḥ.

Post-colophon: likhetaṃ śrīsavāījayapuranagaramadhye |vācakottamaśrīśrīlakṣmīcaṃdradaivajñatatśiṣyaṛṣaladāsena
yatinā | matinā | saṃvat 1813 varṣe mitī phāguṇa vadi 12 dine likhitā sāraṇī śubhā.

Ff. ⟨61v–68v⟩. Table 5 of the *Jagadbhūṣaṇa* (Saturn) for Śaka 1560 to 1677 (= A.D. 1638 to 1755).

Colophon on f. ⟨68v⟩: iti śrīmajjagadbhūṣaṇe śanispaṣṭo saṃpūrṇam.

Post-colophon: saṃvat 1814 varṣe mitī āṣāḍha śudi 4 dine likhitātra || riṇī caiṣā ṛṣaladāsena.

F. ⟨69⟩. Table 19 of the *Jagadbhūṣaṇa* (Rāhu) for 0 to 92 years.

Table 20 of the *Jagadbhūṣaṇa* (Rāhu) for 1 to 27 avadhis.

156. Khasmohor 5420. Ff. 1–7. 11 × 26.5 cm. 12 lines. Verses only. Copied on Wednesday 5 June 1771.

Colophon on f. 7v: iti śrīgaṇakacakracūḍāmaṇibhaṭṭaśrīharajī°
mahārā° paṃcama || 5 ||

After this is written:

> graṃthe smin yad vyalīkaṃ syā jagadbhūṣaṇasaṃjñake |
> sudhībhiḥ śodhanīyaṃ - - - rya puna naḥ || 1
> pūrṇatarka1060śarabhūmi - - - - - dhavalapakṣasamite
> paṃcāṃmītithiyute budhavāre bhūṣaṇaṃ samakarod varadattaḥ || 1 || ||
> saṃpūrṇayaṃ jagadbhūṣaṇasaṃj'nako pathaḥ || 1 ||

Post-colophon: saṃvat 1826 āśāḍe kṛṣṇe tithau 9 budhe lipīkṛtam.

157. Khasmohor 5014. Ff. 1–72. 12 × 28 cm.

F. 1v. Table of the mean motions of the epact, the lord of the year, the Moon, and the lunar anomaly for 20-year periods from Śaka 1520 to 1700 (= A.D. 1598 to 1778); *cf.* table 2 of the Anonymous of 1598 in *SATIUS* 55 a–b.

Table of the mean motions of the same entities for 1 to 20 years; *cf.* table 1 of the Anonymous of 1598 in *SATIUS* 55a.

F. 2. Table of the mean motions of the Moon and the lunar apogee for 1 to 60 ghaṭīs; *cf.* table 14 of the *Jagadbhūṣaṇa*.

F. 2v. Tables 12 and 13 of the *Jagadbhūṣaṇa*.

Tables 10 and 11 of the *Jagadbhūṣaṇa*.

F. 3. Table of the lunar equations, their differences, the lunar velocities, and their differences for 1 to 30 intervals of 3°; *cf.* table 15 of the *Jagadbhūṣaṇa*.

Table of solar longitudes and velocities for 1 to 27 avadhis; *cf.* table 17 of the *Jagadbhūṣaṇa*.

F. 3v. Table 16 of the *Jagadbhūṣaṇa*.

Ff. 4–13v. Table 1 of the *Jagadbhūṣaṇa* (Mars) for Śaka 1560 to 1638 (= A.D. 1638 to 1716).

Ff. 14–25. Table 2 of the *Jagadbhūṣaṇa* (Mercury) for Śaka 1560 to 1605 (= A.D. 1638 to 1683).

Ff. 25v–35v. Table 3 of the *Jagadbhūṣaṇa* (Jupiter) for Śaka 1560 to 1642 (= A.D. 1638 to 1720).

Ff. 36–64. Table 4 of the *Jagadbhūṣaṇa* (Venus) for Śaka 1560 to 1786 (= A.D. 1638 to 1864).

Ff. 64v–71v. Table 5 of the *Jagadbhūṣaṇa* (Saturn) for Śaka 1560 to 1618 (= A.D. 1638 to 1696).

F. 72. Table 19 of the *Jagadbhūṣaṇa*.

Table 20 of the *Jagadbhūṣaṇa*.

Colophon on f. 72v: sāraṇī jagadbhūṣaṇakī.

Indication of price by another scribe: kī. 5|||j.

158. Khasmohor 5050. Ff. 1–4 and 4b–71. 11 × 25 cm.

F. 1v. Table 7 of the *Jagadbhūṣaṇa* (epact) for 0 to 121 years.

F. 2. Table 6 of the *Jagadbhūṣaṇa* (lord of the year) for 0 to 88 years.

Ff. 2–2v. Table 8 of the *Jagadbhūṣaṇa* (Moon) for 0 to 122 years.

F. 2v. Table 9 of the *Jagadbhūṣaṇa* (lunar anomaly) for 0 to 42 years.

F. 3. Table of the mean motions of the Moon and the lunar anomaly for 1 to 60 ghaṭīs; *cf.* table 14 of the *Jagadbhūṣaṇa*.

F. 3v. Table of the mean motions of the Moon and the lunar anomaly for 1 to 27 avadhis; *cf.* tables 10 and 11 of the *Jagadbhūṣaṇa*.

Table of the mean motions of the Moon and the lunar anomaly for 1 to 13 days; *cf.* tables 12 and 13 of the *Jagadbhūṣaṇa*.

F. 4. Table 15 of the *Jagadbhūṣaṇa* (lunar equation) for 0 to 30 intervals of 3°.

Table 17 of the *Jagadbhūṣaṇa* (Sun) for 1 to 27 avadhis

F. 4v. Table 16 of the *Jagadbhūṣaṇa* for 1 to 27 avadhis.

Ff. 4b–13. Table 1 of the *Jagadbhūṣaṇa* (Mars) for 0 to 78 years.

Ff. 14–25. Table 2 of the *Jagadbhūṣaṇa* (Mercury) for 0 to 45 years.
Ff. 25v–35v. Table 3 of the *Jagadbhūṣaṇa* (Jupiter) for 0 to 82 years.
Ff. 36–64. Table 4 of the *Jagadbhūṣaṇa* (Venus) for 0 to 226 years.
Ff. 64v–71v. Table 5 of the *Jagadbhūṣaṇa* (Saturn) for 0 to 58 years.
Colophon on f. 71v: iti śrījagadbhūṣaṇe śanispaṣṭa saṃpūrṇaḥ.

159. Khasmohor 5079. Ff. 1–88. 11.5 × 25.5 cm.
Ff. 1–14. Table 1 of the *Jagadbhūṣaṇa* (Mars) for 0 to 78 years; year 0 = Śaka 1560 (= A.D. 1638).
F. 1. At the bottom is written:

> śāko bhūṣaṭtithibhi 1560 svagajaśvaraḥ cakre sya bhājito
> śeṣaṃ tatpaṃktistho graho bhavet bhaumaspaṣṭaḥ
> gaṇo bhaje cakre 14 vadhe pramāṇaṃ nagādhike lpe
> dviyuk ṛṇasthaṃ evaṃ prasādhyāḥ sphuṭacālakena
> game svarṇaviparītacakre tātkālika raṇaḥ.

Ff. 14v–25v. Table 2 of the *Jagadbhūṣaṇa* (Mercury) for 0 to 45 years; year 0 = Śaka 1560 (= A.D. 1638).
Ff. 26–39v. Table 3 of the *Jagadbhūṣaṇa* (Jupiter) for 1 to 82 years; year 1 = Śaka 1643 (= A.D. 1721).
Ff. 40–77v. Table 4 of the *Jagadbhūṣaṇa* (Venus) for 0 to 227 years; year 0 = Śaka 1560 (= A.D. 1638).
Ff. 78–87v. Table 5 of the *Jagadbhūṣaṇa* (Saturn) for 0 to 58 years; year 0 = Śaka 1560 (= A.D. 1638).
F. 88. Table 19 of the *Jagadbhūṣaṇa* (lunar node) for 0 to 92 years; year 0 = Śaka 1560 (= A.D. 1638) or Śaka 1657 (= A.D. 1735).
Table 20 of the *Jagadbhūṣaṇa* (lunar node) for 1 to 27 avadhis.
F. 88v. Blank except for the title.

160. Khasmohor 5174(m). Ff. 1–5. 5174(b). Ff. 6–69. and 5174(s). F. 69, corrected to 70. 11 × 26 cm.
F. 1v. Table 6 of the *Jagadbhūṣaṇa* (lord of the year) for 0 to 88 years.
Ff. 2–2v. Table 7 of the *Jagadbhūṣaṇa* (epact) for 0 to 121 years.
Ff. 2v–3. Table 8 of the *Jagadbhūṣaṇa* (Moon) for 0 to 121 years.
Ff. 3–3v. Table 9 of the *Jagadbhūṣaṇa* (lunar anomaly) for 0 to 43 years.
F. 3v. Table 10 of the *Jagadbhūṣaṇa* (Moon) for 1 to 27 avadhis.
Ff. 3v–4. Table 11 of the *Jagadbhūṣaṇa* (lunar anomaly) for 1 to 27 avadhis.
F. 4. Table 12 of the *Jagadbhūṣaṇa* (Moon) for 1 to 13 days.
Table 13 of the *Jagadbhūṣaṇa* (lunar anomaly) for 1 to 13 days.
Ff. 4–4v. Table 14 of the *Jagadbhūṣaṇa* (Moon) for 1 to 60 ghaṭikās.
F. 4v. Table 15 of the *Jagadbhūṣaṇa* (Moon) for 1 to 30 sets of 3°.
F. 5. Combination of tables 16, 17, and 18 of the *Jagadbhūṣaṇa* for 1 to 27 avadhis.

Ff. 5v–15v. Table 1 of the *Jagadbhūṣaṇa* (Mars) for 0 to 78 years.
Ff. 16–23v. Table 2 of the *Jagadbhūṣaṇa* (Mercury) for 0 to 45 years.
Ff. 23v–34. Table 3 of the *Jagadbhūṣaṇa* (Jupiter) for 0 to 82 years.
Ff. 34–62. Table 4 of the *Jagadbhūṣaṇa* (Venus) for 0 to 226 years.
Ff. 63–70. Table 5 of the *Jagadbhūṣaṇa* (Saturn) for 0 to 58 years.
F. 70v. Table 19 of the *Jagadbhūṣaṇa* (lunar node) for 0 to 92 years.
Table 20 of the *Jagadbhūṣaṇa* (lunar node) for 1 to 27 avadhis.

161. Khasmohor 5174(l). F. 4. 11 × 26.5 cm. Incomplete.
F. 4. Table 12 of the *Jagadbhūṣaṇa* (Moon) for 1 to 13 days.
Table of lunar equations for 1 to 29; entry for 29 is that for 30 intervals of 3°: 5;2,10°.
Ff. 4–4v. Table 15 of the *Jagadbhūṣaṇa* (Moon) for 0 to 30 intervals of 3°.
F. 4v. Table 19 of the *Jagadbhūṣaṇa* (lunar node); incomplete (for 0 to 26 years).

The **JAGADBHŪṢAṆAKARTTAVYATĀ**. The first verse is:

madhor gatās te khaguṇair vinighnā
yuktās tithī nirgatasaṅkhyakāni ||
śuddhyonitāḥ ṣaṣṭidināgahīnā
gaṇo bhaved abdapateḥ sakāśāt || 1 ||

Manuscript:

162. Khasmohor 5349(c). Ff. ⟨64v–65⟩. 11 × 26 cm.
Colophon on f. ⟨64v⟩: iti jagadbhūṣaṇakarttavyatā samāptā.
F. ⟨65⟩. Table 19 of the *Jagadbhūṣaṇa* (lunar node); incomplete.
F. ⟨65v⟩. Blank.

Tables for 1638 and later years.

Manuscript:

163. Khasmohor 5015(b). Ff. 1–3. 12 × 28 cm.
Ff. 1–1v. Table of the epact for Śaka 1560 to 1706 (= A.D. 1638 to 1784).
Ff. 2–2v. Table of dhruvāṅkas of the Moon and the lunar anomaly for Śaka 1580 to 1680 (= A.D. 1658 to 1758).
Ff. 3–3v. Table of the lord of the year from Śaka 1620 to 1715 (= A.D. 1698 to 1793).

The **GRAHALĀGHAVASĀRIṆĪ** composed by Prema at Madhunagarī in 1656. *CESS* A4, 229a–229b, and A5, 226a. The first verse is:

natvā gaṇeśasya pādāravindaṃ
hariṃ ca sūryapramukhān grahāṃś ca ||
premo grahārthaṃ grahalāghavasya
laghukriyāṃ sāraṇikāṃ prakurve || 1 ||

Manuscripts:

164. Puṇḍarīka jyotiṣa 42. α. F. 1; and β. Ff. 1–14. 13 × 23.5 cm. 8 lines on α. Copied by Ṛddhinātha Vyāsa on Wednesday 5 May 1703 and on *ca.* 6 August 1703.

α. Ff. 1–1v. Verses.

Colophon on f. 1v: iti śrīvaṃśīrasūnumiśrapremaviracitāyāṃ grahalāghavasāriṇyāṃ ślokakramavivaraṇaṃ samāptam.

Post-colophon: saṃ 1760 jyeṣṭhaśukla 1 budhe likhitaṃ vyāsa ṛddhināthenedaṃ pustakam.

β. F. 1. Tables of the mean motion of the Sun for 0 to 29 days, for 30 to 330 days in sets of 30, for 360 to 3960 days in sets of 360, and for 0 to 30 dhruvāṅkas of 4016 days each; the mean longitude of the Sun at dhruvāṅka 0 is that for 18 March 1520: 11^s 19;41,0°.

F. 1v. Similar tables for the Moon; the mean longitude of the Moon at dhruvāṅka 0 is 11^s 19;6,0°.

F. 2. Similar tables for the lunar apogee; the longitude of the lunar apogee at dhruvāṅka 0 is 5^s 17;33,0°.

F. 2v. Similar tables for the lunar node; the longitude of the lunar node at dhruvāṅka 0 is 0^s 27;38,0°.

F. 3. Similar tables for Mars; the mean longitude of Mars at dhruvāṅka 1 is 8^s 11;36,0°; at dhruvāṅka 0, which is omitted, it would be 10^s 7;8,0°.

F. 3v. Similar tables for Mercury's anomaly; the longitude of Mercury's anomaly at dhruvāṅka 1 is 4^s26;6,0°; at dhruvāṅka 0, which is omitted, it would be 9^s 0;3,0°.

F. 4. Similar tables for Jupiter; the mean longitude of Jupiter at dhruvāṅka 0 is 7^s 2;16,0°.

F. 4v. Similar tables for Venus' anomaly; the longitude of Venus' anomaly at dhruvāṅka 0 is 7^s 20;9,0°.

F. 5. Similar tables for Saturn; the mean longitude of Saturn at dhruvāṅka 1 is 1^s 29;39,0°; at dhruvāṅka 0, which is omitted, it would be 9^s 15;20,0°.

F. 5v. Table of the equation of the center of the Sun and the equation of its velocity for 1° to 90°; the maximum is 2;10,45° at 90°.

F. 6. Table of the equation of the center of the Moon and the equation of its velocity for 1° to 90°; the maximum is 5;1,40° at 90°.

F. 6v. Table of the equation of the center of Mars and the equation of its velocity for 1° to 90°; the maximum is 13;0,0° at 90°.

F. 7. Table of the equation of the center of Mercury and the equation of its velocity for 1° to 90°; the maximum is 3;36,0° at 90°.

F. 7v. Table of the equation of the center of Jupiter and the equation of its velocity for 1° to 90°; the maximum is 5;42,0° at 90°.

F. 8. Table of the equation of the center of Venus and the equation of its velocity for 1° to 90°; the maximum is 1;30,0° at 90°.

F. 8v. Table of the equation of the center of Saturn and the equation of its velocity for 1° to 90°; the maximum is 9;18,0° at 90°.

Ff. 9–9v. Table of the śīghra equation of Mars and the equation of its velocity for 1° to 180°; the maximum is 40;0,0° at 135°.

Ff. 10–10v. Table of the śīghra equation of Mercury and the equation of its velocity for 1° to 180°; the maximum is 21;12,0° at 105° to 120°.

Ff. 11–11v. Table of the śīghra equation of Jupiter and the equation of its velocity for 1° to 180°; the maximum is 10;48,0° at 105°.

Ff. 12–12v. Table of the śīghra equation of Venus and the equation of its velocity for 1° to 180°; the maximum is 46;6,0° at 135°.

Ff. 13–13v. Table of the śīghra equation of Saturn and the equation of its velocity for 1° to 180°; the maximum is 5;42,0° at 90° to 105°.

F. 14. Table of elongations from the Sun necessary for the occurrences of the Greek-letter phenomena of Mars, Mercury, Jupiter, Venus, and Saturn.

Ff. 14–14v. Explanatory text.

Date-formula on f. 14v: saṃ 1760 śrā. śu 5 li° ṛddhinātha.

165. Khasmohor 5596. Ff. 1–14. 11.5 × 26.5 cm. 11 lines on f. 1.

Colophon to Prema's verses on f. 1: iti grahalāghavasāraṇī mūlaśl.

The mean motion tables in this manuscript are arranged differently from those in the preceding one. They are for 1 to 33 dhruvāṅkas (called cakras), 1 to 9 days (ekasthāna), 10 to 90 days in groups of 10 (daśasthāna), 100 to 900 days in groups of 100 (śatasthāna), and 1000 to 9000 days in groups of 1000 (sahasrasthāna). These are like the mean motion tables in the *Grahalāghavasāriṇī* ascribed to Nīlakaṇṭha.

F. 1v. Table of the mean motions of the Sun.

F. 2. Table of the mean motions of the Moon.

F. 2v. Table of the motions of the lunar apogee.

F. 3. Table of the mean motions of Mars.

F. 3v. Table of the mean motions of Mercury's anomaly.

F. 4. Table of the mean motions of Jupiter.

F. 4v. Table of the mean motions of Venus' anomaly.

F. 5. Table of the mean motions of Saturn.

F. 5v. Table of the motions of the lunar node.

F. 6. Table of the equation of the center and the equation of its velocity of the Sun.

F. 6v. Table of the equation of the center and the equation of its velocity of the Moon.

F. 7. Table of the equation of the center and the equation of its velocity of Mars.

F. 7v. Table of the equation of the center and the equation of its velocity of Mercury.

F. 8. Table of the equation of the center and the equation of its velocity of Jupiter.

F. 8v. Table of the equation of the center and the equation of its velocity of Venus.

F. 9. Table of the equation of the center and the equation of its velocity of Saturn.

Ff. 9v–10. Table of the śīghra equation and the equation of its velocity of Mars.

Ff. 10v–11. Table of the śīghra equation and the equation of its velocity of Mercury.

Ff. 11v–12. Table of the śīghra equation and the equation of its velocity of Jupiter.

Ff. 12v–13. Table of the śīghra equation and the equation of its velocity of Venus.

Ff. 13v–14. Table of the śīghra equation and the equation of its velocity of Saturn.

F. 14v. Blank.

The **JAGACCANDRIKĀ SĀRIṆĪ** composed by a Jaina; its epoch is 1668. The first verse is:

> sakalasiddhidāyakaḥ saṣararavisamatejavikhyāta ||
> aśvasenanaṃdana namuṃ te vīsamā jinaprāta || 1 ||

Manuscript:

166. Khasmohor 5049(a). Ff. 41–55. 11 × 25 cm. 17 lines.

Colophon on f. 41v, after verse 37 of the instructions: iti śrījagaccaṃdrikā sāraṇīkarttavyatā samāptā.

F. 42. Table of the accumulated adhimāsas, avamas, ahargaṇas, week-days, and lasped years for the beginning of the years Saṃvat 1725, 1744, 1763, 1782, 1801, 1820, 1839, 1858, 1877, 1896, 1915, and 1934 (every 19 years from A.D. 1668 to 1877). The lasped years of the karaṇa in 1668 were 485 so that the epoch was A.D. 1183, the epoch of Bhāskara's *Karaṇakutūhala*.

Table of the accumulated values of the same entities for 1 to 19 years.

Table of the same for 1 to 27 pakṣas beginning with the kṛṣṇapakṣa of Caitra.

F. 42v. Table of the half daylights at Nāgora, Āgorā, Ajamera, and Bīkānera, where the noon equinocitial shadow is 6;0 digits. The Sun's longitudes are sāyana.

Ff. 43–43v. Table of the mean motions of the Sun, the Moon the lunar apogee, Mars, Mercury, Jupiter, Venus, Saturn, and Rāhu for Saṃvat 1725 to 1896 in intervals of 19 years and for 1 to 19 years.

Ff. 44–44v. Table of the mean motions of the same from Caitra to Caitra and the adhimāsa and for 1 to 15 days.

Ff. 45–45v. Table of the mean motions of the same for 0 to 59 ghaṭikās.

F. 46. Table of the equation of the center of the Sun and the equation of its velocity for 0° to 90°; the maximum is 2;10,54° at 90 °.

F. 46v. Table of the equation of the center of the Moon and the equation of its velocity for 0° to 90°; the maximum is 5;2,31° at 90°.

F. 47. Table of the udayāntara of the Sun and the Moon for 0° to 29° horizontal, and 0/6 to 5/11 zodiacal signs vertical.

Ff. 47v–48. Table of the bhaumamandoccaspaṣṭa for 0° to 29° horizontal, 0 to 11 zodiacal signs vertical.

F. 48v. Table of the equation of the center of Mars and the equation of its velocity for 0° to 90°; the maximum is 11;12,53° at 90°.

Ff. 49–49v. Table of the śīghra equation and hypotenuse of Mars for 0° to 180°; the maximum is 41;17,59° at 130°.

F. 50. Table of the equation of the center of Mercury and the equation of its velocity for 0° to 90°; the maximum is 6;3,38° at 90°.

Ff. 50v–51. Table of the śīghra equation and hypotenuse of Mercury for 0° to 180°; the maximum is 21;36,58° at 110°.

F. 51v. Table of the equation of the center of Jupiter and the equation of its velocity for 0° to 90°; the maximum is 5;15,47° at 90°.

Ff. 52–52v. Table of the śīghra equation and hypotenuse of Jupiter for 0° to 180°; the maximum is 10;59,1° at 100°.

F. 53. Table of the equation of the center of Venus and the equation of its velocity for 0° to 90°; the maximum is 1;31,50° at 90°.

Ff. 53v–54. Table of the śīghra equation and hypotenuse of Venus for 0° to 180°; the maximum is 46;30,28° at 140°.

F. 54v. Table of the equation of the center of Saturn and the equation of its velocity for 0° to 90°; the maximum is 7;38,55° at 90°.

Ff. 55–55v. Table of the śīghra equation and hypotenuse of Saturn for 0° to 180°; the maximum is 6;10,7° at 100°.

At the end of each table of the śīghra equation are noted the elongations of the planet from the Sun necessary for the occurrences of its Greek-letter phenomena.

Colophon on f. 55v: iti jagaccaṃdrikā sāraṇī samāptā.

A table of the lord of the year for every year from Śaka 1600 to 1755 (= A.D. 1678 to 1833) following the Brāhmapakṣa.

Manuscript:

167. Khasmohor 5082(d). F. 1. 11 × 24 cm.

On f. 1v, at the top, is written: brahmapakṣe saṃvat 1735 śā 1660 ta ārabhya varṣamuṣo nibījabdapaḥ prativarṣe kṣime kṣepaka 1|15|31|17|17. The lord of the year in Śaka 1600 (= A.D. 1678) was 1;17,5,30.

The **ANONYMOUS OF 1679**.

Manuscript:

168. Khasmohor 5060. α ff. 1–27; β ff. 1–24; γ ff. 1–24; δ ff. 1–2; ϵ ff. 1–3; ζ ff. 1–4; η ff. 1–26; and θ ff. 1–24. 16 × 30 cm.

For each planet twelve dhruvas are given in which are the planet's true longitudes, differences, and divisors for 0 to 365 days. Each dhruva begins at a different longitude, and after each dhruva are given the longitude at which each Greek-letter phanomenon occurs.

α. Ff. 1–27v. Dhruvas 1 to 12 for Venus.

β. Ff.1–24v. Dhruvas 1 to 12 for Jupiter.

γ. Ff. 1–24v. Dhruvas 1 to 12 for Mercury.

δ. Ff. 1–2v. Table of the initial values for the lord of the year, the epact, the Moon, the lunar anomaly, Mars, Mercury, Jupiter, Venus, Saturn, and Rāhu for the years from Śaka 1601 to 1700 (= A.D. 1679 to 1778).

ϵ. Ff. 1–3v. Table of the true longitudes of the Sun, the solar velocity, the length of daylight (not filled in), the half-equation of daylight (not filled in), and the traikya for 0 to 365 days.

ζ. Ff. 1–2v. Table of the mean longitudes of the Moon and of the lunar anomaly for 0 to 365 days.

Ff. 3–4v. Table of the lunar equations, their differences, and the Moon's velocities for $0^s\ 0^\circ$ to $11^s\ 29^\circ$.

η. Ff. 1–26v. Dhruvas 1 to 12 of Mars.

θ. Ff. 1–24v. Dhruvas 1 to 12 of Saturn.

The **ANONYMOUS OF 1718**, following the Brāhmapakṣa.

Manuscript:

169. Khasmohor 5254(h). F. 1. 11.5 × 26.5 cm.

F. 1. Table of the lord of the year, the epact, and the lunar anomaly for Śaka 1640, 1660, 1680, 1700, 1720, 1740, 1760, 1780, and 1800 (= A.D. 1718 to 1878), and for 1 to 20 years.

F. 1v. Table of the longitudes of the Moon for the same times.

The **ANONYMOUS OF 1726**, following the Brāhmapakṣa.

Manuscript:

170. Khasmohor 5082(f). 1 f. 11 × 25.5 cm.

F. ⟨1⟩. Table of the epact for the beginning of every year from Śaka 1648 to 1695 (= A.D. 1726 to 1773).

Table of the lord of the year for the same.

F. ⟨1v⟩. Table of longitudes of the Moon for the beginnings of the same years.

Table of the lunar anomaly for the beginning of every year from Śaka 1648 to 1697 (= A.D. 1726 to 1775).

The **PAÑCĀṄGASĀRIṆĪ** composed in 30 verses by Kevalarāma at Jayapura in 1735. *CESS* A6. The first verse is:

varadāyakaṃ nikhilavighnavārakaṃ
gaṇanāyakaṃ ruciracaṃdrabhālakam ||
aghahārakaṃ śaraṇagaikapālakaṃ
praṇamāmi taṃ smararahasya bālakam || 1 ||

Manuscripts:

171. Puṇḍarīka jyotiṣa 48. Ff. 1–2 and ⟨3⟩. 12 × 26 cm. 12 lines. With marginalia. Copied by Gokula on Wednesday 3 April 1793.

Colophon of f. ⟨3⟩: iti śrījyotiṣarāyakevalarāmakṛtā paṃcāṃgasāraṇī.

Post-colophon: likhitā gokulena madhau asitapakṣe ṣaṣṭyāṃ iṃdusute saṃvat 1850.

F. ⟨3v⟩. Blank.

172. Puṇḍarīka jyotiṣa 47. Ff. 1–3. 13 × 27.5 cm. 10 lines. With marginalia. Copied by Gokulanātha (the same scribe as Gokula of the manuscrit Puṇḍarīka jyotiṣa 48) on Sunday 28 April 1793.

Colophon on f. 3v: iti śrījyotiṣarāyakevalarāmakṛtā paṃcāṃgasāraṇī.

Post-colophon: likhitā gokulanāthena vaiśākhakṛṣṇe dvitīyāyāṃ ravau saṃ 1850.

A **SŪRYASIDDHĀNTASĀRIṆĪ** whose epoch is 1743.

Manuscript:

173. Khasmohor 5174(n). Ff. 1–4. 11 × 26 cm.

F. 1. Table of the values of the lord of the year, the epact, the mean Moon, the lunar apogee, and the lunar node at the beginnings of Śaka 1665, 1685, 1705, 1725, 1745, and 1765 (= A.D. 1743 to 1843) and for 1 to 20 years.

F. 1v. Table of the mean longitudes of Mars, Mercury, Jupiter, Venus, and Saturn at the same times.

Ff. 2–2v. Table of the mean motions of each of the above mentioned for 1 to 12 months, 1 to 4 periods of 6 days, and for 1 to 5 days.

Ff. 3–3v. Table of the equations of the centers of the Sun. the Moon, Mars, Mercury, Jupiter, Venus, Saturn, and their differences for 0 to 30 intervals of 3°; their maxima, at 90°, are 2;11° for the Sun, 5;2,41° for the Moon, 11;37,49° for Mars, 4;28° for Mercury, 5;6° for Jupiter, 1;45° for Venus, and 7;40° for Saturn.

Ff. 4–4v. Table of the śīghra equations of Mars, Mercury, Jupiter, Venus, and Saturn for 1 to 30 intervals 6°; the maxima are 40;16° for Mars at 22=132°, 21;31° for Mercury at 19=114°, 11;31° for Jupiter at 17=102°, 46;21° for Venus at 23=138°, and 6;22° for Saturn at 16=96°.

F. 4v. Table of the longitudes of the apogees of the Sun, Mars, Mercury, Jupiter, Venus, and Saturn.

Table of the elongations from the Sun of the five star-planets necessary for the occurrences of their first and second stations.

Table of the mean daily motions expressed in minutes and seconds of the Sun, the Moon, the lunar apogee, the lunar node, Mars, Mercury's śīghra, Jupiter, Venus' śīghra, and Saturn.

A tithi, nakṣatra, and yoga table whose epoch is 1747.

Manuscript

174. Khasmohor 5254(b). Ff. 1–23; and 1 blank f. 11.5 × 26.5 cm.

F. 1v. Tables of tithis, kendras, and an unidentified function for Śaka 1669 to 1797 (= A.D. 1747 to 1875) at intervals of 16 years, and for 1 to 16 years.

Tables of nakṣatras and yogas for Śaka 1669 to 1789 (= A.D. 1747 to 1867) at intervals of 24 years, and for 1 to 24 years.

F. 2. Table of the beginnings of the quarters of the 28 nakṣatras.

F. 2v. Table of the days of the Sun's entry into each of the 12 zodiacal signs and into each of the 29(!) nakṣatras.

Ff. 3–7v. Table of tithis for 0 to 27 horizontal and 0 to 27 vertical.

Ff. 8–8v. Table of tithis for 1 to 28 horizontal and 1 to 13 months vertical.

Ff. 9–9v. Table of equations of the tithis for 0 to 27 horizontal and 1 to 13 vertical.

Ff. 10–14v. Table of nakṣatras for 0 to 27 horizontal and 0 to 27 vertical.

Ff. 15–15v. Table of nakṣatras for 1 to 28 horizontal and 1 to 13 months vertical.

Ff. 16–16v. Table of equations of the nakṣatras for 0 to 27 horizontal and 1 to 13 vertical.

Ff. 17–21v. Table of yogas for 0 to 29 horizontal and 1 to 29 vertical.

Ff. 22–22v. Table of yogas for 1 to 30 horizontal and 1 to 13 months vertical.

Ff. 23–23v. Table of equations of the yogas for 1 to 29 horizontal and 1 to 13 vertical.

The **SŪRYASIDDHĀNTASĀRIṆĪ** or **GRAHASPAṢṬASĀRIṆĪ** in 21 verses composed by Candrāyaṇa at Mulatāna in 1748. See *CESS* A3, 46a, and A5, 109a. The first verse is:

natvārkaṃ sūryasiddhāntasaṃsiddhagrahasāraṇī ||
candrāyaṇena kriyate gaṇitakleśahāriṇī ||

Manuscripts:

175. Khasmohor 5250. 1 f.; and 1 blank f. 11 × 26 cm. 17 lines. Copied by Tejabhāna on Thursday 11 May 1769.
Colophon on f. 1: iti grahaspaṣṭasā(raṇī).
Post-colophon entirely in margin: (saṃvat 1826 vaiśākha sudi 6 gurau tejabhānena lipteyaṃ).
F. 1v. Blank.

176. Khasmohor 5495(a). F. 1. 10.5 × 26.5 cm. 10–13 lines. Copied by Sūrjabhāna on Wednesday 30 August 1769.
Colophon on f. 1v: iti śrīgrahaspaṣṭasāraṇī.
Post-colophon: saṃ 1826 bhādrapavadavadi 14 budhavāre sūrjabhānena likhitā sāraṇī.

177. Khasmohor 5495(b). 1 f. 10.5 × 26.5 cm. 3 lines. Incomplete. On the verso is patra 4 of Saturn's śīghraphala.

The **TITHIKALPAVṚKṢA** composed by Candrāyaṇa in 11 verses. See *CESS* A3, 46a, and A5, 109a. The first verse is:

natvā gaṇeśasya padāravindaṃ
śrīkṛṣṇadevasya guroś ca mūrdhnā |
budhvā hi sauraṃ tithikalpavṛkṣaṃ
pañcāṅgapatrāśubhavāyu kurve ||

Manuscript:

178. Khasmohor 5251. 1 f. 11 × 26 cm. 9 lines. With marginal notes. Copied by Sūrjabhāna on Friday 1 September 1769.
Colophon on f. 1v: iti śrīcandrāyaṇadaivajñaviracitas tithikalpavṛkṣaḥ.
Post-colophon: saṃ° 1826 bhādrasudi 1 śukre likhitaṃ sūrjabhānena.

Table unfilled in of Mercury's equation of the center according to the *Karaṇakutūhala* of Bhāskara and dated Śaka 1674 (= A.D. 1752).

Manuscript:

179. Khasmohor 5174(c). F. 3. 12.5 × 26 cm.
Obverse and recto: skeleton of a table. The title of the table on the reverse is: vudhādīnā maṃdaphalakhaṃḍāni.

At the top of the reverse is written: śakaḥ 1674 gatābde piṃḍaḥ 569. Śaka 1674 diminished by 596 is Śaka 1105 (= A.D. 1183), the epoch of Bhāskara's *Karaṇakutūhala*.

The **LAGHUTITHICINTĀMAṆI** composed by Nandarāma Miśra at Kāmavana in 1777. See *CESS* A5, 158a. The first verse is:

haripadam abhivandya viśvahetos
tithimukhasiddhipaṭaṃ tanomi samyak |
kṛtam api makarandapūrvavijñair
na milati samprati kāladeśabhedāt ||

Manuscript:

180. Khasmohor 5574. Ff. 1–8. 12 × 26.5 cm. 12 lines. Instructions only.
Colophon on f. 8v: iti śrīmirśranaṃdarāmaviracite (corr. to viracito) laghuciṃtāmaṇiḥ.

A **GRAHALĀGHAVAKARTTAVYATĀ** in Hindī. It begins: tatrādau mūlasūtrānusāreṇa śuddāhargaṇaḥ sādhyaḥ.

Manuscript:

181. Khasmohor 5504(a). Ff. 1 and 11 (f. 11 also numbered 59 on the verso). 10.5 × 24.5 cm. 17 lines. Incomplete. Copied at Saṅgrāmapura on Tuesday 27 March 1716.

F. 1v ends: jo keṃdrarāśi ghāmai 6|7|8|9|10(|11).

F. 11 begins with a table of Saturn's śīghra equation with the entries for arguments from 151 to 180; for 151 the equation is 3;12,0°.

Colophon after this table: iti śrīgrahalāghavakī sāraṇī saṃpūrṇā.

Post-colophon: saṃvat 1773 varṣe | vaiśākhakṛṣṇapratipattithau | bhaumavāre likhitaṃ | śrīsaṃgrāmapuramadhye.

Below this are 5 lines of text and another colophon: iti grahalāghavasāraṇīkarttavyatā saṃpūrṇā.

There follows a table of the motions in 7 days and the yearly motions of the Sun, the Moon, the lunar apogee, the lunar node, Mars, Mercury, Jupiter, Venus, and Saturn, and the lord of the year.

Table of the lord of the year for 1 to 120 years with the parameter of Gaṇeśa's *Grahalāghava*.

Manuscript:

182. Khasmohor 5174(u). F. 2. The yearly parameter is 1;15,31,30 days.

A set of calendrical and planetary tables

Manuscript:

183. Khasmohor 5353(b). Ff. 3–10. 11 × 25 cm. Incomplete.

Ff. 3–3v. Tables of annual motions, epoch positions, dhruvāṅkas without Rāmabījas, Rāmabījas, and dhruvāṅkas with Rāmabījas of the epact, the lord of the year, Mars, Mercury, Jupiter, Venus, Saturn, and the lunar node. Unfortunately, no numerical entries were recorded for this manuscript.

F. 4. Tables of the epact for 0 to 9, 10 to 90 in steps of 10, and 100 to 900 in steps of 100 years.

Ff. 4–4v. Tables of the lord of the year according to the same arrangement.

F. 4v. Tables of Mars according to the same arrangement.

Tables of Mercury according to the same arrangement.

Ff. 4v–5. Tables of Jupiter according to the same arrangement.

F. 5. Tables of Venus according to the same arrangement.

Tables of Saturn according to the same arrangement.

Tables of the lunar node according to the same arrangement.

No descriptions were recorded of ff. 5v–7 and 8–9.

F. 7v. Tables of the longitudes of the lunar node for 0 to 27 avadhis.

Ff. 9v–10v. Table of koṣṭhakākṣepāḥ for 0 to 365 days.

A set of planetary tables consisting of a "saurabha" of each planet for 0 to 59.

Manuscript:

184. Khasmohor 5174(i). Ff. ⟨1⟩ and 2–3. 11.5 × 25 cm.

F. ⟨1⟩. Ravisaurabha.

⟨Candrasaurabha⟩.

F. ⟨1v⟩. Bhaumasaurabha.

F. 2. Budhasaurabha.

F. 2v. Gurusaurabha.

F. 3. Bhṛgu⟨saurabha⟩.

F. 3v. Śani⟨saurabha⟩.

For the Sun there is 1 column, perhaps relating to true longitude; for the Moon 2 columns, perhaps relating to mean longitude and the equation; and for the five star-planets 3 columns each, perhaps relating to mean longitude and the 2 equations.

A set of tables of the planets' longitudes set up like those of Mahādeva's *Mahādevī*.

Manuscript:

185. Khasmohor 5253(b). Ff. 17–69. 11 × 26 cm. Incomplete.

F. 17. Mars for N=58–59.
Ff. 17v–47. Mercury for N=0–59.
Ff. 47v–62. Jupiter for N=0–59.
Ff. 62v–69v. Venus for N=0–29.

Table of the longitudes of the lunar node for 0 to 100 years and for years 61 to 77.

Manuscript:

186. Khasmohor 5174(w). F. 1. 11 × 26 cm. The epoch longitude is 8^s 22;52,40°.

Tithi, nakṣatra, and yoga tables.

Manuscript:

187. Khasmohor 5520. Ff. 2 and 4. 11.5 × 26 cm. Incomplete.
F. 2. Blank.
F. 2v. Tithi table with 4 variables; for 0 to 26 horizontally.
F. 4. Price mark: kī ऽ।.
F. 4v. Table of two functions—vārādi and vallī—from 63 to 125 horizontal.

A table in Hindī relating to tithis 8 and 0 of the kṛṣṇapakṣa and tithis 8 and 15 of the śuklapakṣa of āsāḍha, srāvaṇa, bhadravā, āsu, kātī, magasira, poha, māha, phāguṇa, caitra, vaïsāṣa, jeṭha, āsāḍha, śrāvaṇa, bhādravā, āsū, kātī, magasira, posa, māha phāguṇa, caitra, vaisāṣa, and jeṭha.

Manuscript:

188. Khasmohor 5082(g). 1 f. 25.5 × 10.5 cm. Torn at the top.

A sāyanalagnasāriṇī for a latitude where the noon equinoctial shadow is 6;6 (ϕ=26;57°).

Manuscript:

189. Puṇḍarīka jyotiṣa 46. 1 blank f. and ff. 1–2. 12 × 26 cm. Copied by Gokulanātha Jośī on Friday 14, 21, 28 November, or 5 or 12 December 1788.
Ff. 1–2v. The table is for 0 to 29° horizontal and 0 to 11 zodiacal signs vertical.
Date-formula on f. 2v: likhitā iyaṃ jośī gokulanāthena āgrahāyaṇike māsi śukre saṃvat 1845 śāke 1710.

A sāyanalagnapatra for Lavapura, where the noon equinoctial shadow is 7;30 (ϕ=32;0°) and the deśāntara is 35 yojanas.

Manuscript:

190. Khasmohor 5174(r). 1 f. 11 × 26.5 cm.

A nirayanalagnasāraṇī for Mūlatrāṇa. The entry is 0;0,0 at 11^s 10° so that the precession amounts to 20°; this indicates a date of about 1840.

Manuscript:

191. Khasmohor 5254(d). Ff. 1–1v. 11.5 × 26.5 cm.

A nirayanalagnapatra. The entry for o^s 0° is 2;11.

Manuscript:

192. Khasmohor 5349(d). 1 f. 11 × 26 cm.

A nirayanalagnasāraṇī for Jayapura, where the noon equinoctial shadow is 6;6 (ϕ=26;57°); the precession is 21;21°, which indicates a date of about 1900.

Manuscript:

193. Puṇḍarīka jyotiṣa 43(a). 2 ff. 10 × 21.5 cm.
Ff. ⟨1–1v⟩. Nirayanalagnasāriṇī.
Ff. ⟨2–2v⟩. Table of the lords of the year, the nakṣatras, and the tithis for 1 to 100 years. The yearly parameter for the lord of the year is that of the *Grahalāghava*, 1;15,31,30 days.
Below this is written:

tigmāṃśuśukraśaniśukrasurejyacaṃdrāḥ
śaśāṃkabhaumaśanibhaumasurejyacaṃdrāḥ ||
devejyaśītakiraṇeṃdujabhūmiputrāḥ
sūryośanaḥśanisitārkikujejyacaṃdrāḥ ||

A nirayanalagnasāraṇī for a locality where the noon equinoctial shadow is 6;6 (ϕ=26;57°).

Manuscript:

194. Puṇḍarīka jyotiṣa 44(e). Ff. 1–2. 10 × 21.5 cm.

A table of the ascendents and midheavens for 0 to 29° horizontal and four zodiacal signs (Cancer to Libra?) vertical.

Manuscript:

195. Khasmohor 5082(b). 1 f. 10.5 × 25.5 cm. Incomplete.
First sign, 0°: ascendent 3^s 6;3,58, midheaven 11^s 29;57,36°.
Fourth sign, 28°: ascendent 6^s 23;6,48°, midheaven 3^s 24;7,26° (the ascendent for fourth sign, 29°, is not entered).

A table of the fourth bhāva.

Manuscript:

196. Khasmohor 5174(q). F. 1. 11 × 26.5 cm.

A table for the fourth and tenth bhāvas. The precession is stated to be 20°, which indicates a date of about 1840.

Manuscript:

197. Khasmohor 5504(b). 1 f. 11 × 25.5 cm.

A table of the tenth bhāva.

Manuscript:

198. Khasmohor 5254(f). 1 f. 11.5 × 26.5 cm.

A table of the cusps of the tenth bhāva as each of the 28 nakṣatras is in the ascendent. Composed by Viśveśvara in about 1790/1800 at Jayapura for Pratāpasiṃha. See *CESS* A5, 700b–701b.

Manuscript:

199. Puṇḍarīka jyotiṣa 44(g). 2 ff. 10 × 21.5 cm.
Ff. ⟨1–1v⟩. A text in 4 verses. It begins:

arkāṃśaiḥ pramitaṃ śaśāṃkabhavanaṃ dāsre khamadhyasthite
yāmye tryaśvimitair atho taraṇibhaṃ tarkāṃśakair vahnibhe ||
śailābjaiś ca khabhe 'tha śūnyalavayuk vānyā mṛge rudrabhe
vedāṃśair aditau gajāśvibhir atho puṣye tulāsaṃjñakaṃ || 1 ||

It ends on f. ⟨1v⟩:

itthaṃ dṛksamatāṃ vilokya katavāṃ(?) śrīmatpratāpājñayā
dāsrādau ca khamadhyage jayapure viśveśvaro veśmakaṃ || 4 ||

See Puṇḍarīka jyotiṣa 9, f. A (manuscript 234).

F. ⟨1v⟩. Table of the cusps of the tenth bhāva when each of the 28 nakṣatras is in the ascendent.

Ff. ⟨2–2v⟩ blank.

A table of the length of daylight at Mūlatrāṇa.

Manuscript:

200. Khasmohor 5174(d). F. 1. 10.5 × 25 cm. The maximum is 34;56 ghaṭikās when the Sun is at Gemini 9°(sic!), the minimum is 25;2 ghaṭikās at Sagittarius 10°. The precession is 20°, indicating a date of about 1840.

A table of half the length of daylight.

Manuscript:

201. Khasmohor 5254(c). F. 1. 11.5 × 26.5 cm. The maximum is 17;29 ghaṭikās when the Sun is at Gemini 10°, the minimum is 12;31ghaṭikās at Sagittarius 10°. The scheme, then, is virtually identical to that in Khasmohor 5174(d) (manuscript 200).

A table of Sines for 1 ° to 90°; R=60.

Manuscript:

202. Khasmohor 5481. 1 f. 15 × 32.5 cm.

A table of the shadows of a gnomon 12 digits in height when the Sun has altitudes of 1° to 90°; this is a table of Cotangents with R=12.

Manuscript:

203. Khasmohor 5480. 1 f. 15 × 33 cm.

A table of the altitude of the Sun for 1 to 15 ghaṭikās on the equinoctial days and the lengths of the shadows cast by a 12-digit gnomon. The noon equinoctial shadow is 6;6 (ϕ=26;57°).

Manuscript:

204. Puṇḍarīka jyotiṣa 44(f). Ff. 1–⟨2⟩. 10 × 21.5 cm.

A table of the reversed shadow for 0° to 90°. This is equivalent to a Tangent table with R=3438.

Manuscript:

205. Khasmohor 5479. 1 f. 14.5 × 28 cm.

Tables of unidentified functions.

Manuscripts:

206. Khasmohor 5174(a). F. 1. 11 × 26 cm. Table of a function with differences for an argument of 0 to 59. The entry for 1 and 59 is 6,33; the maximum is 100,0, the entry for 30.

207. Khasmohor 5174(f). 1 f. 11 × 27 cm. The argument was 1 to 12 vertical and 0 to 27 horizontal; only the first 3 lines on the recto are filled in. The entries range from +0,27 to -2,40; the entry for 27 in line 3 is 0,0.

208. Khasmohor 5174(o). 1 f. 10.5 × 24.5 cm.

F. ⟨1⟩. Table for 1 to 60; the entries do not have a common difference. The entry for 1 is 11,14,6,44,13,48; that for 60 is 12,36,4,2,40,48.

F. ⟨1v⟩. A similar table, but entries are given only for arguments from 1 to 30. The entry for 1 is 10,48,18,40,12; that for 30 is 24,9,20,6,0.

E. ECLIPSES

The **DHĪKOṬIDAKARAṆA** composed by Śrīpati in 1038. See *CESS* A6. Edited by K. S. Shukla, Lacknow 1969. The first verse is:

natvā bhavānīcaraṇāravindaṃ
śivaṃ gaṇeśaṃ druhiṇaṃ girāṃ ca |
harīṃ guruṃ sadgrahaṇaṃ raveś ca
candrasya vakṣye sugamaprakāram ||

Manuscripts:

209. Puṇḍarīka jyotiṣa 27. Ff. 1–3. 11.5 × 23.5 cm. 9 lines. Copied by Śrīnātha on Thursday 30 August 1632.
F. 1v begins with 1,2a (p. 6,9): caṃdrāṃganaṃdo961naśako.
Colophon on f. 3: iti raviparvādhikāraḥ.
Post-colophon: saṃvat 1689 āśvanavadi 10 daśamyāṃ gurau likhitaṃ śrīnātha.
F. 3v. Blank.

210. Khasmohor 5129. Ff. 1–⟨2⟩. 11 × 24.5 cm. 9 lines. Presented as one unified text with 19 verses.
F. 1 begins in 1,2a (p. 6,9): caṃdrāṃganaṃdo961naśako.
Colophon on f. ⟨2⟩: iti śrīśrīpatibhaṭṭaviracite dvikoṭināmā karanaṃ samāptam iti.
F. ⟨2v⟩. Blank.

211. Khasmohor 5300. Ff. 1–3. Dimensions not recorded. 9 lines.
F. 1v begins in 1,2a (p. 6,9): caṃdrāṃganaṃdo961naśako.
Colophon on ff. 2v–3: iti śrīśrīpatibhaṭavira[f. 3]citaṃ caṃdrasūryayor grahaṇajñānaṃ.
Price on f. 3v: kīmati S|||.

The **CHĀDIKANIRṆAYA** composed by Kṛṣṇa in *ca.* 1600. See *CESS* A2, 51a; A4, 58a; and A5, 49a–49b. The first verse is:

guror viṣṇunāmnaḥ padāmbhogayugmaṃ
praṇamyāntarāyātadhīvṛddhikāri |
niśānāthaghasreśayoś chādakasya
grahe nirṇayaṃ vaktum icchāṃ karomi ||

Manuscript:

212. Puṇḍarīka jyotiṣa 7. Ff. 1–6. 10 × 23 cm. 12 lines. Formerly property of Viśveśvara, the son of Rāmeśvara.

Colophon on f. 6v: iti śrīsakalagaṇakasārvabhaumaballālagaṇakātmajakṛṣṇadaivajñaviracitaḥ chādakanirṇayaḥ saṃpūrṇaḥ.

Ownership note on f. 1:

> rāmeśvaratanūjasya viśveśvaravipaścitaḥ ||
> chādakanirṇayapustaṃ jagaj jānātv idaṃ śubhaṃ || 1 ||

Tables for computing eclipses.

Manuscript:

213. Puṇḍarīka jyotiṣa 44(c). Ff. 1–14. 10.5 × 21.5 cm.

Ff. 1–3v. Table of the corrected longitudes of the Sun for each degree of its mean longitude, its daily motions, the differences in its equation, and its caras.

Ff. 4–6v. A similar table for the Moon with the differences in the daily motions substituted for the Sun's caras.

Ff. 7–9v. Table of deflections (valanāṅghrayaḥ) for lunar eclipses in the western hemisphere and for solar eclipses in the eastern hemisphere.

Ff. 10–12v. Table of deflections (valanāṅghrayaḥ) for lunar eclipses in the eastern hemisphere and solar eclipses in the western hemisphere.

F. 13. Table of lunar latitudes.

Ff. 13v–14v. Table of dṛkkarmāṇi.

A set of eclipse tables.

Manuscript:

214. Puṇḍarīka jyotiṣa 44(d). Ff. 1–2. 10.5 × 21.5 cm.

Ff. 1–2. Table of the diameters of the lunar disk measured in digits for daily motions from 722;20 (=12;2,20°) to 858;50 (=14;18,50°).

F. 2. Table of the diameters of the solar disk measured in digits for daily motions from 0;56,53° to 1;1,22°

F. 2v. Table of lunar latitudes measured in aṅgulas.

A **GRAHAṆAMĀLĀ** giving details of solar and lunar eclipses between Wednesday 15 śuklapakṣa of Pauṣa in Saṃ 1796 = 16 January 1740 and Monday 15 śuklapakṣa of Kārttika in Saṃ 1846 = 16 November 1789.

Manuscript:

215. Puṇḍarīka jyotiṣa 272. Ff. 1–5. 10 × 21 cm. 13/14 lines. Only the versos of the folia written upon.

On f. 2 a second scribe has written:

śrīrāmajīsahāya || śrīvāgīśvarī jayati ||
vāmadevo mahādevo virūpākṣas trilocanaḥ ||
kṛśānu.

Observation reports of eclipses.

Manuscripts:

216. Khasmohor 4876. 1 f. 119 × 42.5 cm. 45 lines. Torn. With a diagram of a lunar eclipse in blue and yellow. The first verse is:

sa jayati sindhuravadano
devo yatpādapaṃkajasmaraṇaṃ |
vāsaramaṇir iva tamasāṃ
rāśiṃ nāśayati vighnānām || 1 ||

It computes a lunar eclipse that was observed at 52;46 ghaṭikās of Monday 15 śuklapakṣa of Vaiśākha in Saṃ. ⟨1818⟩, Śaka 1683 = 18 May 1761.

217. Khasmohor 5178. 1 f. 94 × 22 cm.
Recto: computation according to the *Brahmatulya* of a lunar eclipse at 38;59 ghaṭikās on Friday 14 śuklapakṣa of Caitra in Saṃ. 1829, Śaka 1694 = 17 April 1772.
Verso: computation according to the Brahmapakṣa of a solar eclipse at 9;28 ghaṭikās on Tuesday 30 kṛṣṇapakṣa of Caitra in Saṃ. 1829, Śaka 1694 = 23 March 1773.

218. Khasmohor 5246. 3 ff. Each has a diagram of its eclipse.
F. ⟨1⟩. 36 × 25.5 cm. 11 lines. Computation according to the *Brahmatulya* of a lunar eclipse on Thursday 15 śuklapakṣa of Āśvina in Saṃ. 1830, Śaka 1695 = 30 September 1773.
F. ⟨2⟩. 26.5 × 21.5 cm. 13 lines. Computation according to the *Brahmatulya* of a lunar eclipse on Wednesday 15 śuklapakṣa of Māgha in Saṃ. 1831, Śaka 1696 = 15 February 1775.
F. ⟨3⟩. 26.5 × 21.5 cm. 5 lines. Computation according to the *Brahmatulya* of a solar eclipse on Tuesday 30 kṛṣṇapakṣa of Caitra in Saṃ. 1829, Śaka 1694 = 23 March 1773.

219. Khasmohor 4877. 1 f. 61 × 43 cm. In two columns. With a diagram of the eclipse.
Column 1: computation according to the *Grahalāghava* of a lunar eclipse at 8;54 ghaṭikās on Monday 14 śuklapakṣa of Māgha in Saṃ. 1916, Śaka 1781 = 5 February 1860.

Column 2: on what measures to take to avoid dire consequences.

220. Khasmohor 4878. 1 f. 60.5 × 44 cm. The first verse is:

> yaṃ brahmadevāṃtavido vadaṃti
> paraṃ pradhānaṃ puruṣaṃ tathānye ||
> viśvodgateṣ kāraṇam īśvaraṃ vā
> tasmai namo vighnavināśanāya || 1 ||

Copied by Choṭelāla.

Computation according to the *Brahmatulya* of a lunar eclipse at 3;20 ghaṭikās on Tuesday 15 śuklapakṣa of Māgha in Saṃ. 1916, Śaka 1781 = 6 February 1860. At the end is the verse:

> śāstrānusāreṇa mayā lipikṛtam agre samyag
> īśvaro śrīśivo vetty alam || choṭelālena viduṣā likhyate parvvapa-
> trikā ||

F. STAR CHARTS

A star chart made for Mahārāja Mādhavasiṃha (who ruled Jayapura from 1751 to 1768) according to a verse in its middle:

> jayati yantram idaṃ jhaṭiti sphuṭaṃ
> diśati darśanato 'sya niśāgatam ||
> tanunṛpāspadapatnisukhānvitaṃ
> nṛpatimādhavasiṃhavinirmitam ||

It is accompanied by two set of tables.

Manuscript:

221. Khasmohor 1257. There are three parts.

(2). The chart. 1 piece of starched cloth. 73.5 × 76.5 cm. Attached to the center is a red silk thread with a red and green silk tassel, which acts as an index.

There are four concentric circles. The outermost, labelled ghaṭijñānārtham, is numbered at equal intervals from 1 to 60 ghaṭikās, and each ghaṭikā is divided into 6 equal parts each equivalent to 10 vighaṭikās or palas.

The next circle, labelled sūryādīnāṃ saptamabhāvajñānaṃ, is divided into 12 zodiacal signs, with zero at 18;50 ghaṭikās (=113 time-degrees). Also units of 6°are marked and single degrees. It is used for determining the descendent.

The next circle, labelled daśamacaturthayor jñānam, is marked similarly to the second circle, except that zero is at 3;10 ghaṭikās (=19 time-degrees). It is used for determining the cusps of the fourth and tenth places.

The innermost circle, labelled lagnajñānaṃ, is marked similarly to the second and third circles, except that zero is at 47;30 ghaṭikās (=285 time-degrees). It is used for determining the ascendent.

Below the central verse are two columns listing the altitudes in degrees (unnatāṃśas) and meridian-distances in ghaṭikās (nataghaṭīs) of 21 stars for each integer number of ghaṭikās and the maximum number of ghaṭikās and palas between the horizon and meridian. The twenty-one stars and the maximum meridian-distance of each are (the pala numbers are difficult to decipher and may be erroneous):

Aśvinī:	16;41.
Rohiṇī:	16;24.
Ārdrā:	15;30.
Punarvasu:	17;41.
Maghā:	16;21.
Pūrvā Phālgunī:	16;40.
Uttarā Phālgunī:	16;24.
Citrā:	14;10.
Svāti:	16;44.

Viśākhā:	13;30.
Anurādhā:	13;4.
Jyeṣṭhā:	12;48.
Abhijit:	18;40.
Śravaṇa:	15;45.
Pūrvā Bhādrapadā:	17;29.
Uttarā Bhādrapadā:	17;47.
Prajāpati:	20;6.
Brahmahṛdaya:	20;16.
Lubdhaka:	13;41.
Matsyamukha:	12;4.
Kukkuṭapuccha:	20;0.

On the outer periphery of the outermost circle are marked the lines of the risings (udaya), culminations (madhya), and settings (asta) of these stars:

	Udaya	Madhya	Asta
Aśvinī	47;28	4;10	21;52
Bharaṇī		6;25	
Kṛttikā		8;46	
Rohiṇī	54;38	10;55	27;24
Mṛgaśīrṣa		13;24	
Ārdrā	58;36	14;11	29;48
Punarvasu	1;40	18;42	36;20
Puṣya		22;02	
Āśleṣā		22;48	
Maghā	8;52	24;49	40;58
Pūrvā Phālgunī	10;34	27;32	44;28
Uttarā Phālgunī	12;38	29;0	
Hasta		30;42	
Citrā	18;48	32;58	47;08
Svāti	18;14	35;12	52;08
Viśākhā	23;52	36;32	50;08
Anurādhā	26;24	39;24	52;28
Jyeṣṭhā	27;58	40;34	53;14
Mūla		43;18	
Pūrvāṣādhā		45;12	
Uttarāṣādhā		45;22	
Abhijit	27;12	46;10	5;08
Uttarāṣādhā (sic)		46;40	
Śravana	33;20	49;04	4;50
Dhanisthā		51;02	
Śatabhiśa		56;32	
Pūrvā Bhādrapada	39;38	57;06	14;36
Uttarā Bhādrapada	42;08	59;48	17;22
Prajāpati	53;12	14;12	34;10

Brahmahṛdaya	52;08	12;24	32;40
Lubdhaka	2;52	16;24	29;58
Matsyamukha	44;42	56;48	8;52
Kukkuṭapuccha	31;22	51;22	11;22

(1). 12ff. (a: 6 ff.; b: ff. 1–6). 20 × 30.5 cm. 14 lines. Designed to be hinged to the left. Each page in each half is for one zodiacal sign. Each has an argument from 0° to 29° horizontal.

a. To the left of each page are the names of the relevant nakṣatras and the times of their risings and culminations as the Sun is in each degree, to the right is a column of ascendents in degrees, minutes, and seconds.

b. To the left of each page are the names of the relevant nakṣatras and the times of their risings as the Sun is in each degree.

SŪRYAPATTRAS for the 12 zodiacal signs and pattras for 15 stars which are not nakṣatras. The Sūryapattras have as horizontal argument 0° to 29° of solar longitude in each zodiacal sign; the vertical argument consists of the nakṣatras that are visible during the time that the Sun is in that zodiacal sign. The time that each nakṣatra becomes visible, expressed in ghaṭikās, palas, and vipalas, on each night that the Sun has a given longitude is entered in the table. The tables for non-nakṣatra stars also have as horizontal argument 0° to 29°, but as vertical argument the appropriate zodiacal signs. Again, the times of the star's becoming visible on each night are recorded.

Manuscripts:

222. Puṇḍarīka jyotiṣa 20(a). 1 blank folium and f. 8. 13 × 26.5 cm. Incomplete.

Ff. 8–8v. Sūryapattra for Scorpius. The nakṣatras are Śravaṇa through Āśleṣā. The times are for ghaṭikās and palas only.

223. Puṇḍarīka jyotiṣa 43(b). Ff. 1–12 and 12b–20. 10 × 21.5 cm.
Ff. 1–1v. Sūryapattra for Aries. Punarvasu through Uttarāṣāḍhā.
Ff. 2–2v. Sūryapattra for Taurus. Puṣya through Śravana.
Ff. 3–3v. Sūryapattra for Gemini. Pūrvaphālgunī through Pūrvabhādrapadā.
Ff. 4–4v. Sūryapattra for Cancer. Citrā through Aśvinī.
Ff. 5–5v. Sūryapattra for Leo. Anurādhā through Rohiṇī.
Ff. 6–6v. Sūryapattra for Virgo. Mūla through Ārdrā.
Ff. 7–7v. Sūryapattra for Libra. Pūrvāṣāḍhā through Punarvasu.
Ff. 8–8v. Sūryapattra for Scorpius. Śravaṇa through Maghā.
Ff. 9–9v. Sūryapattra for Sagittarius. Śatabhiṣaj through Uttaraphālgunī.
Ff. 10–11v. Sūryapattra for Capricorn. Uttarabhādrapadā through Viśākhā.
Ff. 12–12v. Sūryapattra for Aquarius. Aśvinī through Jyeṣṭhā.
Ff. 12b–12bv. Sūryapattra for Pisces. Rohiṇī through Mūla.

F. 13. Pattra for Brahmahṛdaya. Virgo through Pisces.
F. 13v. Pattra for Lubdhaka. Libra through Aries.
F. 14. Pattra for Lubdhakabandhu. Libra through Aries.
F. 14v. Pattra for Mahāpuruṣa. Scorpius through Taurus.
F. 15. Pattra for Siṃhahṛdaya. Sagittarius through Gemini.
F. 15v. Pattra for Pretaśiraḥ. Leo through Aquarius.
F. 16. Pattra for Mithunavāmahasta. Virgo through Pisces.
F. 16v. Pattra for Nadyantaka. Leo through Aquarius.
F. 17. Pattra for Viśākhāmātṛmaṇḍala. Aquarius through Leo.
F. 17v. Pattra for Aśvamukha. Gemini through Sagittarius.
F. 18. Pattra for Kukudapuccha. Taurus through Scorpius.
F. 18v. Pattra for Agni. Virgo through Pisces.
F. 19. Pattra for Sarpadhāriśiraḥ. Pisces through Virgo.
F. 19v. Pattra for Marīci. Capricorn through Cancer.
F. 20. Pattra for Samudrapakṣin. Cancer through Capricorn.
F. 20v. A text in 4 lines describing the construction and use of these tables. It begins: sveṣṭasāyanaspaṣṭasūryarāśipatrāt tadaṃśeṣṭanakṣatrasaṃpātajakoṣṭako grahyaḥ. It ends: khamadhyagatanakṣatraspaṣṭaghaṭya iṣṭāṃśās tad⟨d⟩aśamaghaṭībhir ūnāḥ śeṣaṃ tasminn a⟨ṃ⟩śe rātrigataṃ bhavati.

A treatise on computing star charts. It begins: iṣṭanakṣatrasya nāḍīvalayākhyayaṃtre prācyāḥ paścimā vā nataghaṭikā grāhyāḥ || tās tatspaṣṭaghaṭīṣu ūnā vā yutāḥ kāryās.

Manuscript:

224. Puṇḍarīka jyotiṣa 45. Ff. 1–2. 10 × 21.5 cm. 9 lines. There are 4 prakāras: 1 untitled; 2 spaṣṭaghaṭyānaya⟨na⟩; 3 ghaprādyas (ghaṭyādyas?)tad⟨d⟩aśamānayana; and 4 ghaprādi (ghaṭyādi?) prātar daśamānayana.

F. 2v ends: tadrāśyaṃśasaṃpātajakoṣṭakaḥ kalādyanupātena yuktaḥ spaṣṭo bhavati | sa eva dyaprādi (ghaṭyādi?) prātar daśamaṃ bhavati.

A description of the special relationships between stars, giving the color and magnitude of some. It begins: teṣām ekanakṣatraṃ kṛttikāsaṃjñam asti. The stars described are: Kṛttikā, Brahamahṛdaya, Hasta=Rohiṇī, Jabbāra, Khaḍga, Kaṭau, Mithunavāmahasta, Ārdrā, Vāmapāda, Mṛgaśiras, Ākāśagaṅgā, Lubdhaka, Bandhulubdhaka, Punarvasu, Bālaśiras, Maghā, Pharahada (Yavanabhāṣayā), Sarpa, Pūrvaphālgunī, Uttaraphālgunī, Siṃhapuccha, Citrā, Svāti, Hasta, Kanyāvāmahasta, Viśākhāmātṛmaṇḍala, Vṛścikahṛdaya=Jyeṣṭhā, Abhijit, Śravaṇa, Sarpadharapuruṣaśiras, Kukkuṭapuccha, and Katikara.

Manuscript:

225. Puṇḍarīka jyotiṣa 26. Ff. 1–4. 10 × 21.5 cm. 9 lines. Formerly property of Dhaneśvara.

F. 4 ends in line 3 and at the beginning of the line 4: evaṃ mayā viṃśati nakṣatrāṇi kathitāni || ebhya unnatāṃśajñānaṃ bhavati.

F. 4v. Blank.

Ownership note on f. 1: pustakam idaṃ dhaneśvarasya śubhaṃ.

G. GEOGRAPHICAL TABLES

A list of some of the place names mentioned in the *Bhāgavata*, *Brahmāṇḍapurāṇa*, *Brahmapurāṇa*, *Brahmavaivartta*, *Garuḍapurāṇa*, *Harivaṃśa*, *Kūrmapurāṇa*, *Laghunāradīya*, *Liṅgapurāṇa*, *Mahābhārata*, *Mārkaṇḍeyapurāṇa*, *Matsyapurāṇa*, *Padmapurāṇa*, *Rāmāyaṇa*, *Śivapurāṇa*, *Skandapurāṇa*, *Vāmanapurāṇa*, *Vārāha*, *Vāyupurāṇa*, and *Viṣṇupurāṇa*.

Manuscript:

226. Khasmohor 2954. Ff. 1–5. 58 × 56 cm. Only rectos inscribed.
At the bottom of f. 5 is written: paṃcapatre(ṣu) navaśatapaṃcadaśa deśāḥ 915.

A table of localities with their noon equinoctial shadows, their equations of daylight (sometimes), and their oblique ascensions of the zodiacal signs. Composed before the foundation of Jayapura in 1727.

Manuscript:

227. Khasmohor 5082(a). F. 1. 11 × 25.5 cm.
The cities and their noon equinoctial shadows are:

āmbera 6|0
paṭṭaṇā 5|45
cītroḍa 5|30
nāgora 5|54
ajamera 5|52, corrected by a later scribe to 6|9
lāhora 7|30
buṃdī 5|30
ṣaṃbhāita 4|51
bījāpura 3|30
mathurā 6|28
jālora 5|11
kurukṣetra ⟨6⟩|53
laṅkodayalagnaṃ
gaṃgarā 6|10
dilī 6|20
meḍato 5|56
rājanagare 5|6
laṃkā
jaimārаṇi 5|53

In the margin a later scribe has written: udayapure sāyamurānagare.
The remainder of the recto and the verso have an undistinguished text on astronomy and astrology.

A much fuller table entitled *Palabhājñāna*. Composed after 1727 since it includes Ja⟨ya⟩pura.

Manuscript:

228. Puṇḍarīka jyotiṣa 50. Ff. 1–4. 10 × 21.5 cm. 9 lines. Formerly property of Dhaneśvara Pauṇḍarīka. A complete transcription follows:

[f.1v] śrīgaṇeśāya namaḥ || atha palabhā ||

laṃkāpuryāṃ || akṣabhā 0|0|| akṣakarṇa 0|0|| carakhaṃḍakaḥ || 00||00||00|| paramadina ||30||00||

kāṃtīnagaryāṃ || akṣabhā ||3||23|| akṣāṃśa ||15||43|| akṣakarṇa 12||28|| carakhaṃḍaka ||34||27||11|| paramadina ||32||22||

anāgūdī || bījānagara || akṣabhā ||4|| akṣāṃśā ||18||33||24||akṣakarṇa ||12||14|| carakhaṃḍaka ||40||32||13|| paramadina ||32||50||

anāttadaṃveradara || akṣabhā ||4||15|| akṣāṃśa ||19||38|41|| akṣakarṇa ||12 ||28|| carakhaṃḍaka ||42||34||14|| paramadina ||33||0|| laṃkāyojana ||146||

bhṛgukacha bhavokṣa || akṣabhā ||4||51|| akṣāṃśa || 22||10||30|| akṣakarṇa ||12||56|| carakhaṃḍaka ||48||39||16|| paramadina ||33||26|| laṃkāyojana ||192||

devagiri || akṣabhā ||4|25|| akṣāṃśa ||20||19||36|| karṇa ||12|43|| carakhaṃḍaka ||44||38||15|| paramadina ||33||8|| laṃkāyojana ||167||

tilaṃgā || akṣabhā ||3||55⟨||⟩ [f.2] akṣāṃśa || 18||24|| paramadina ||32||46|| karṇa ||12||37|| carakhaṃḍaka ||39||31|| 13| laṃkāyojana ||146||

avaṃtī || akṣabhā ||5||9|| akṣāṃśa ||23||15||9|| karṇa ||13||5|| carakhaṃḍaka ||41||41||57|| paramadina ||33||38|| laṃkāyojana ||206||

ajamera | akṣabhā ||6|| akṣāṃśa ||26||32||9|| karṇa ||13||25|| carakhaṃḍaka ||60||48|20|| paramadina ||34||16|| laṃkāyojana ||247||

āśrīpattanī ⟨|| a⟩kṣabhā ||5||35|| akṣāṃśa ||24||14|| akṣakarṇa ||13||25|| carakhaṃḍaka ||56||44||18|| paramadina ||33||56|| laṃkāyojana ||227||

citrakūta || akṣabhā ||5||30|| akṣāṃśa ||24||34||33|| akṣakarṇa ||13||13|| carakhaṃḍaka ||55||44||18|| paramadina ||33||54| laṃkāyojana ||223||

gargarād ⟨||⟩ akṣabhā ||5[0]||20|| akṣāṃśa ||24||00|| karṇa ||23||27|| carakhaṃḍaka ||53||43||18|| paramadina ||33||48|| laṃkāyojana ||215||

nāgora ⟨||⟩ 'kṣabhā ||5||58| akṣāṃśa ||26||00|| karṇa ||13||42|| carakhaṃḍaka ||59||47||19|| paramadina ||34||12|| laṃkāyojana ||246||

[f. 2v] sīt⟨ā⟩pura [||] uttare ⟨||⟩ akṣabhā ||4(|)40|| akṣāṃśa ||21(|)13(|)0|| karṇa ||12||52|| carakhaṃḍaka ||47||27||15|| paramadina |33||54|| laṃkāyojana ||180||

somanāthapattana || akṣabhā ||5⟨||⟩6|| akṣāṃśa ||23||9||37|| akṣakarṇa ||13||42|| carakhaṃḍakaḥ ||51||41||17|| paramadina ||33||38|| laṃkāyojana ||204||

meḍherū anahalavāḍu [||] pātana gujarāta || akṣabhā ||5||20|| akṣāṃśāḥ ||24|0|| akṣakarṇa ||13||10|| carakhaṃḍaka ||53||43||18|| paramadina ||33||48|| laṃkāyojana ||218||

tihuṇagiri || akṣabhā ||6|0|| akṣāṃśāḥ ||26||32|| akṣakarṇa ||13|42|| carakhaṃḍaka ||51|41|17|| paramadina ||34||38||16|| laṃkāyojana ||247||

abhoraū || akṣabhā ||(7)||13⟨||⟩ akṣāṃśāḥ ||31|0|0|| akṣakarṇa ||14|0|| carakhaṃḍaka ||72||58||24|| paramadina ||35||8|| laṃkāyojana ||318||

yoginīpura dillī || akṣabhā ||6||34|| akṣāṃśāḥ ||28||34|| akṣakarṇa ||13(|)40|| carakhaṃḍaka ||65||53||22|| paramadina ||34(|)4|| laṃkāyojana ||272||

hāsī ḍhīlīsamīpe || akṣabhā ||6(|)35|| akṣāṃśāḥ ||28||56⟨||⟩ [f. 3] akṣakarṇa ||13||44|| carakhaṃḍaka ||66|53|22|| paramadina ||34||42|| laṃkāyojana ||278||

gopāvala [|| a]kāliṃjara || akṣabhā ||5||45|| carakhaṃḍaka ||57||46||29||
lākherī ||5||30|| cara ||55||44||18||
vurahānahana || akṣabhā ||4||30|| carakhaṃḍaka ||54||36||15||
sīhānade ||akṣabhā ||7||
dilī ||6|20||
āgaro ||6||10||
vīdhānagara ||2|51||
bīkānera ||7|0||
varodā ||4||52|
valada ||8|10||
virāṭ ||6||27|
bharoṃca ||4|55||
bhṛgukacha ||4||58||
māṃdoṃ ||4||47||
mathurā ||6|2||
murora ||5||46||
mulatāna ||6(|)4||
mithilā ||6(|)0||
jodhapura ||5|58||
japura ||6(|)6||
bharatapura ||6||
romeśvara ||1|30||
rājamahala ||5|35||
rohitaka ||6|37||
rājagaḍha ||5|30⟨||⟩
raṇathaṃbhora ||5||34||
somanātha ||3|6|
saṃbhala ||6|21|
lāhora ||7||30|
lavaṃgapura ||5|35||
lasavatī ||5(|)0⟨||⟩
lakhaṇaü ||5||45||
sīhora ||5|00|
sāgarapura ||5||52|
stabhatīrtha ||4|51||
geḍhadeśa ||5|10||
gayā ||5|50⟨||⟩
gokula ||6|00||
gagharāṭ ||5|25||
cola ||4|00|
cadelī ||5(|)5|
jālora |5|13||

jalālāvāda ||8|00||
[f. 3v] chagopa ||3||00||
ṭhadoṃ ||5||36||
ugūrapura ||5|(3)⟨||⟩
ṭākā ||6|20||
tājapura ||5(|)42||
tailaṃga ||4[5]|40|
thānapura ||6|30||
dirasa |6||20|
dvārikā ||5|5|
radaurakā ||4||25||
dolatāvāda ||4|15|
dvārāvatī ||6||5|
devagiri ||4||25|
tilaṃga ||3|55||
vuṭānapura ||4||30||
avaṃtikā |5(|)9||
ajamera ||6(|)00||
śrīpattana ||5|35||
viguḍha |5|30|
gargarāṭ ||5⟨||⟩20|
gaḍhā ||4||32||
gahora ||5||47|
ayodhyā ||5|52|
ajamera ||5||6||
ruhamadāvāda |6⟨||⟩00||
avaṃtī ||5||25||
amarakoṭa ||4||36|
ahamadānagara ||6|12|
ābhānora ||6|45|
āpaseṭatā ⟨||⟩6||20|
iṃya |5(|)2|
utkala |5[(1)5]|45|
udayapura ||5(|)35|
oṭa ||4(|)6|
oṭhachā ||5||45[||2|]
kāśmīra ||7|52|
kābila ||8||25||
kālipī ||5|45|
kāṃtī ||3||13||
kovana ||3||15||
kanoja ||6(|)20|
kurukṣetra ||6|36⟨||⟩
kaṃpilā ||6|16|
kedāra ||7(|)1||
khaṃbhāyana ||6||45||

gaṃgāsāgara ||6||57||
golakuṃḍā ||3|45||
gvāliyara ||5||52||
vabūnagara ||6||00||
nāgora ||5|58||
śīdapura ||4||40||
gujarāta ||5||20||
pāṭaṇa ||3|36||
hari[f. 4]dvāra ||6||36||
hastināpura ||6|21||
hedarāvāda ||4||00||
samasāvāda ||6(|)21||
sūrata ||4||47|
dohara ||5(|)5||
deśāha ||5||4||
śadhigāma ||4||40||
dābhola ||5||0|
damanapura ||5||45||
dholapura ||5||2|
dhāmonī ||5||0⟨||⟩
niramadā ||5(|)5||
nāgora ||5||51||
nagarakoṭa ||7||40|
nāsaketa ||5|4|
narmadā ||4||47||
nadanagara ||5(|)5⟨||⟩
kopāla ||6⟨||⟩20||
naimiṣa ||5||45||
prayāga ||5||45||
vāṇarasī ||5||45||
paṭaṇā ||5||45||
peśora ||8⟨||⟩00⟨||⟩
papari ||4||30||
prakāśa ||4||51||
pāṃḍava ||4(|)1||
puruṣottama purī ||5||56||
panātaveda ||4|51||
vedara ||3|45⟨||⟩
bahlīgapura ||4||30|
būdī ||5||30|
bījāpura ||3||6||
śrīnagara ||6||40|

jodhapura || akṣabhā ||5||58|| akṣāṃśa ||26||54 | karṇa ||13|24| carakhaṃḍaka ||59||47||20|| paramadina ||34||12| laṃkāyojana ||245 ⟨||⟩

thāneśvara kurukṣetra ⟨||⟩ akṣabhā ||6||50|| akṣāṃśā ||30||10| karṇa ||13|51|| [a]carakhaṃḍaka ||62||55|23||

kumāü ||akṣabhā ||6||44||
F. 4v. Blank.
Ownership-note on f. 1: palabhādipustaka pauṃḍarīka || dhaneśvarasya.

H. ASTRONOMICAL INSTRUMENTS

The **YANTRARĀJĀGAMA** composed by Mahendra Sūri for Fīrūz Shāh, the Tughluq Sulṭān from 1351 to 1388, in about 1370. See *CESS* A4, 393b–395a, and A5, 296b–297a. Edited with the ṭīkā of Malayendu by K.K. Raikva, Mumbayī 1936. The first verse is:

śrīsarvajñapadāmbujaṃ hṛdi parāmṛśya prabhāvapradaṃ
śrīmantaṃ madanākhyasūrisuguruṃ kalyāṇakalpadrumam |
lokānāṃ hitakāmyayā prakurute sadyantrarājāgamaṃ
nānābhedayutaṃ camatkṛtikaraṃ sūrir mahendrābhidhaḥ ||

Manuscripts:

229. Khasmohor 4958. ff. 1–46. 11.5 × 27.5 cm. 9/10 lines. Ff. 1 and 46 torn. Papaer old and torn. With the ṭīkā of Malayendu. With some marginalia. Copied by Kṛṣṇa on Friday 5 March 1596.

Colophon on ff. 45v–46: samāptā yaṃtrarājādhyāyasya ṭīkā kṛtir iyaṃ śrīmadbṛhadgachamaṃḍana[f. 46]śrīmahendrasūriśiṣyaśrīmalayacaṃdrasūriṇā viracitā.

After this are written 30 verses on various astronomical problems. The first is:

nāgāṃbhodhiyamāḥ surāḥ sitabudhau pūrvoditau 248|33 tiṣṭhataḥ
prācyām astamitau mataṃgajanagā 78 vṛṃdārakā 33 vāsarāḥ ||
vāruṇyām uditau divākaravaśāt tau śīghrabhuktī sadā
tāvaṃttye vadinānyadṛśyavapu[śau]ṣau tulyāni digbhūmitaiḥ || 1 ||

The last is:

nāḍyaś care paṃcadaśo janānyaś
caikaṃ dinaṃ ṣaṣṭighaṭīpramāṇaṃ ||
niśāpi tāvatpramitaiś ca bhūyāt
saumyaṃ tv ajāder vaṇijaś ca yāmyaṃ || 30 ||

After this is the post-colophon:

vājibhūtithi1517mite śakakāle
śuklaphālgunaharau bhṛguvāre ||
ravipuṣye ||
kṛṣṇasaṃjñagaṇakaḥ svaparārthaṃ
yaṃtrarājam alikhad vividhārthaṃ || 1 ||
kiṃ vānena dhanena vājkaribhiḥ prājyena rājyena kiṃ
kiṃ vā putrakalatramitrapaśubhir dehe(na) gehena kiṃ ||
hitvaitat kṣaṇabhaṃ gur⟨u⟩ṃ sa padire sarvaṃ mano dūrataḥ
svātmārthaṃ guruvākyato bhaja bhaja śrīpārvatīvallabhaṃ || 1 ||

230. Khasmohor 4974. Ff. 1–15. 11.5 × 26 cm. 10 lines. Paper darkened in places. With occasional pūrvamātras. With the ṭīkā of Malayendu. Copied by Śrīdatta, the son of Śaṅkaradāsa Vyāsa of the Śrīmālajñāti, a resident of Jodhapura, on Saturday, 11 January 1617 during the reign of Mahārājādhirāja Sūryasiṃha, who ruled Marwar from 1595 to 1620.

Colophon on f. 15v: iti śrīyantrarājāgame spaṣṭādhyāyaḥ paṃcamaḥ.

Post-colophon: saṃvat 1673 varṣe śāke 1538 pravarttamāne pauṣamāse śuklapakṣe paurṇimāsyāṃ tithau śanivāre likhito yaṃ yantrarājaḥ || jodhapuranagare vāstavyaṃ mahārājādhirājamahārā śrīsūryasiharājye śrīmālajñātī vyāśaśaṃkaradāsātmajena śrīdattena likhitam ātmapaṭhanārtham iti.

After this is written: pratyaśuddhatvād aśuddhaṃ kiṃcid asti madīyo na doṣaḥ. yādṛśaṃ dṛṣṭaṃ tādṛśaṃ likhitaṃ.

Price on f. 1: kī° ≡.

231. Khasmohor 4973. Ff. 1–2, ⟨3⟩, and 4–42. 11 × 24.5 cm. 15–17 lines. With marginalia and an occasional pūrvamātra. With the vyākhyā of Malayendu Sūri. Copied by Śrīdatta, a resident of Śubhaṭapura, the son of Śaṅkara Vyāsa, the son of Puruṣottama Vyāsa of the Śrīmālījñāti, at Śuddhadantī on Sunday 2 November 1617.

Colophon on f. 42: samāptā yaṃtrarājādhyāyasya ṭīkā || kṛti śrīmadvṛhadgache śrīmalayacaṃdrasūriṇā.

Post-colophon: saṃvat 1674 varṣe śāke 1539 pravarttamāne śuklapakṣe pūrṇi kārttikamāse pūrṇimāyāṃ tithau raviravivāsare vyā° śrīmālījñātīyavyāsapurṣottamasutavyāsaśaṃkara tatputraśrīdattaliṣitaṃ ātmapaṭhanārthaṃ || śubhaṭapuraṭamadhye vāstavyaṃ || śuddhadaṃtīmadhye likhitaṃ.

Price on f. 1: kī° |ɿ.

232. Khasmohor 5469. Ff. 1–59. 13 × 31 cm. 11 lines. F. 20v blank with no break in text. With the vyākhyā of Malayendu Sūri. Copied from manuscript 229 by Tulārāma at the command of Mahārājādhirāja Jayasiṃha on Monday 23 September 1706.

Colophon on f. 58v: samāptā yaṃtrarājādhyāyasya ṭīkā kṛtir iyaṃ śrīmadbṛhadgacchamaṃḍanaṃ (śrīmaheṃdrasūriśiṣyaśrīmalayacaṃdrasūriṇā viracitā).

Post-colophon on ff. 58v–59: saṃvat 1763 varṣe āśvinakṛṣṇa 13 some [f. 59] śrīmanmahārājādhirājaśrījayasiṃhadevajīkasyājñayā likhitaṃ idaṃ tulārāmeṇa.

On f. 59 are written the 30 verses found on f. 46 of manuscript 229 followed by the beginning of that manuscript's post-colophon:

> vājibhūtithi1517mitai śakakāle
> śuklaphālgunaharau ravipuṣpe.

Price on f. 1: kī ¶||| ≡.

233. Khasmohor 4975. Ff. 1–16. 12 × 26.5 cm. 8/9 lines. With some marginalia. With Malayendu Sūri's final verse at the end. Copied by Tulārāma at the command of Mahārājādhirāja Jayasiṃha on Saturday 16 November 1706.

Colophon on f. 16: iti śrīyaṃtrarājā samāptaḥ.

Post-colophon: śrīmanmahārājādhirājajīśrījayasiṃhadevajīkasyājñayā saṃ 1763 varṣe mārgaśīrṣakṛṣṇa 8 śanau likhitam idaṃ tulārāmeṇa.

Price on ff. 1 and 16v: kī |=ɹ.

234. Puṇḍarīka jyotiṣa 9. Ff. ⟨1⟩, 2–25-1, 25, 25-2, and 26–45; and f. ⟨A⟩. 12.5 × 27.5 cm. 12/13 lines. F. 25-1v blank. With the vyākhyā of Malayendu Sūri. With marginalia. Copied from manuscript 229 by Gopīnātha, the son of Rāmeśvara Vyāsa, for Jāgeśvarajīka on Sunday 14 December 1788.

F. ⟨1v⟩. Table of the Sines for 1° to 90° with R=60. The entries have sexagesimal fractions with 4 to 6 places.

Ff. 2–2v. Table of the Sines with differences for 1° to 90° with R=3600. This is the table found on pp. 2–4.

Ff. 2v–3. Table of the Versines with differences for 1° to 90° with R=3600. This is the table on pp. 4–6.

F. 3 begins in 1,5a (p. 2, 11): yat syāl lavādhaḥ kalikādikaṃ ta(d).

F. 24v ends in the vyākhyā on 1,67 (p. 46, 27–28): anyac ca kā.

F. 25-1 is a śodhapatra written by a second scribe, who also copied f. 25-2 and f. ⟨A⟩. It begins in the vyākhyā on 1,67 (p. 46, 28): lacakraparidhicakrāṃśānāṃ.

F. 25-1 ends in the vyākhyā on 1,68–69 (p. 49, 16–17):chāyākaṇau (prasādhyau).

F. 25-1v. Blank.

F. 25 begins: athaiṣāṃ pratyaṃśaṃ koṣṭakeṣu nyāso yathā. There follows the table on pp. 47–49.

F. 25v ends with the table on pp. 49–51.

Ff. 25-2–25-2v is another śodhapatra copied by the second scribe. Its contents are virtually identical with those of ff. 25–25v.

F. 26 begins with 1,71a (p. 51,21): abhūd.

Colophon on ff. 45–45v: samāptā yaṃtrarājādhyāyasya ṭīkā kṛtir iyaṃ śrīmadbṛhadgachamaṃḍanaśrīmahemdrasūriśiṣyaśrīmala[f. 45v]yacaṃdrasūriṇā viracitā.

After this is written: śrīrāmaḥ śaraṇaṃ mameti paramo maṃtro yam aṣṭākṣaraḥ.

There follow the 30 verses, the post-colophon, and the final verse found on f. 46 of manuscript 229. Then: śrīviśveśvaraḥ prīṇātv anena śrameṇa.

Post-colophon on f. 45v: likhitam idaṃ pustakaṃ vyāsopanāmakarāmeśvarasyātmajagopīnāthena śrījāgeśvarajīkasya paṭhanārthaṃ ca || saṃvat 1845 śāke 1710 pravarttamāne pauṣavadi 1 ravivāre.

On f. ⟨A⟩ the second scribe has written: atha valayayaṃtraṃ ||

vṛttārddhaṃ dhṛti18bhāgaṃ ekaikasyāpi paṃcabhir guṇitaṃ |
tasyāṃśake caturthe chidre sūcyagrabhāgena || 1 ||

bhāgā nirekā guṇasaṃguṇās te
dinārddhabhaktā dinaśeṣanāḍyaḥ ||

iti valayayaṃtraṃ ||

atha cāvukayaṃtraṃ ||

śaṃku⟨ḥ⟩ prakalp⟨y⟩o ⟨'r⟩kamitāṃgulo ⟨'⟩tra
tanmānato vuddhimatā ca vṛddhiḥ ||
yaṣṭi⟨ḥ⟩ prakalp⟨y⟩ākṛtisaṃguṇā vā
syāt kalpanā sadgurusaṃpradāyāt || 1 ||

This is Hema, *Kaśāyantra* 2.

śaṃkuprabhāgraṃ spṛśatīha ya[ṃ]tra[ṃ]
taccihnataḥ ⟨'⟩sre[a] khalu tadpramāṇe |
jñeyaṃ[b] dyumāne ⟨'tra⟩ tu[c] yanmitāṃkas
tattulyasaṃkhyā ghaṭikā krameṇa || 2 ||

This is a variant of Hema, *Kaśāyantra* 14.

iti cāvukayaṃtraṃ ||

a. sve. b. neyaṃ. c. ca.

varttamānaśākena saha etasyāṃ1527taraṃ kṛtvā
caturddhā sthāpyaṃ paścād anena 1|15|30|31 guṇyaṃ
kṣepayutaṃ tahavīla.

1	4	7	2	7	3	6	1	3	4	6	7
34	13	29	42	8	20	6	20	7	34	9	41
30	53	38	16	48	32	24	50	40	33	40	30

Then a third scribe has written 4 verses identical to those found on f. ⟨1–1v⟩ of manuscript 199.

F. ⟨Av⟩. Blank.

235. Khasmohor 5206(b). F. 5. 23.5 × 10 cm.

F. 5–5v. Table of the shadow cast by a 7-digit gnomon as the altitude of the Sun increases from 1°to 90°. This is the bottom half of the table on pp. 49–51.

Colophon on f. 5v: samāpto yaṃ grathaḥ.

The **YANTRARĀJĀGAMAVYĀKHYĀ** composed by Malayendu Sūri

in about 1382. See *CESS* A4, 363a–364b, and A5, 282b–283a. Edited by K. K. Raikva, Mumbayī 1936. The firsr verse is:

praṇamya sarvajñapadāravindaṃ
sūrer mahendrasya padāmbujaṃ ca |
tanoti taḍgumphitayantrarāja-
granthasya ṭīkāṃ malayendusūriḥ ||

See manuscripts 229–232 and 234, and *cf.* manuscript 233.

The **DHRUVABHRAMAYANTRA** composed by Padmanābha as part of his *Yantraratnāvalī* in about 1400. See *CESS* A4, 170b–171b, and A5, 205b. The first verse is:

śrīnarmadānugrahalabdhajanmanaḥ
pādāravindaṃ janakasya sadguroḥ |
natvā triyāmāsamayādibodhakaṃ
dhruvabhramaṃ yantravaraṃ bravīmy atha ||

Manuscripts:

236. Puṇḍarīka jyotiṣa 11. Ff. 1–6. 12.5 × 27 cm. Tripāṭha. With the ṭīkā by Padmanābha. Copied by Gopīnātha, the son of Rāmeśvara Vyāsa, on Thursday 5 November 1789. Formerly property of Viśveśvara, the son of Rāmeśevara.

The ṭīkā begins on f. 1v: atha dhruvabhramākhyayaṃtrādhikāraḥ prārabhyate ||

Colophon of mūla on f. 6v: iti dhruvabhramākhyayaṃtraṃ samāptam.

Post-colophon: saṃvat 1846 śāke 17011 mārgaśirṣakṛṣṇe 3 guruvāsare likhitam idaṃ pustakaṃ vyāsopanāmakarāmeśvarasyātmajagopīnāthena.

The ṭīkā ends on f. 6v: athāparoditaḥ dhruvakenāpi na proktaṃ tad etat kautukāt kṛtaṃ.

F. 6v. Blank.

Ownership note on f. 1:

rāmeśvaratanūjasya viśveśvaravipaścitaḥ ||
dhruvabhramākhyayaṃtrasya saṭīkasya hi pustakaṃ |

237. Khasmohor 5206(a). Ff. 1–5. 10 × 23.5 cm. 9 lines.

On ff. 1v–2 is a text on determining the length of nighttime in 10 verses. The first of these is: dhruvamatsyopari[a] rātryānayanaṃ

nijodayair meṣabhato[b] vilagnaṃ ||
bhūbhūta51nāḍījanitaṃ[c] ⟨tu⟩ yasmāt ||

ūrddhvasthapu⟨c⟩che[d] sati tad vilagnaṃ
rūpākṣi(21)nāḍījam[e] adha⟨ḥ⟩sthapu⟨c⟩che || 1 ||

a. °matsyaupari. b. mekhabhato. c. °nāḍījanitaṃ. d. urdvastha°. e. rupākṣi°.

The last verse is:

tadvyāsasakte mukhato ⟨+ +⟩ re
yathā tathā dhāryam idaṃ karābhyāṃ
pu⟨c⟩chaṃ tu vai vorddh⟨va⟩talāṃkato ⟨'⟩ṃśā
natonnatā keṃdragalaṃbam uktā || 10 ||

F. 2v begins, after śrīgaṇeśāya namaḥ, in 1a: śrīnarmadānugrahalabdhaṃ. Colophon on f. 5v: iti śrīpadmanābhaviracitaṃ dhruvabhramayaṃtraṃ samāptaṃ.

The **DHRUVABHRAMAYANTRAṬĪKĀ** composed by Padmanābha in about 1400. See *CESS* A4, 171b–172a, and A5, 205b. It begins: atha dhruvabhramākhyayaṃtrādhikāraḥ prārabhyate || tatrāpi nirvighnaparisamāptyartham abhīṣṭadevatānamaskārarūpaṃ maṅgalam ācaratīndravaṃśena.
See manuscript 236.

The **YANTRACINTĀMAṆI**, composed by Cakradhara, presumably in the sixteenth century. See *CESS* A3, 36b–37b; A4, 88a; and A5, 103b–104a. The first verse is:

natvā phalāptyai pramathādhināthaṃ
raviṃ guror aṅghryaravindayugmam |
yantraṃ pravakṣye gaṇitānapekṣam
āśuprabodhaṃ samayādikānām ||

Manuscript:

238. Khasmohor 4954. Ff. 1–4. 12 × 26.5 cm. 8 lines. With some marginalia. Copied by Tulārāma at the command of Mahārājādhirāja Jayasiṃha on Thursday 14 November 1706.
Colophon on f. 4v: iti śrīvāsudevasutataṃtrajñasiṃhaśrīcakradharaviracitā (corrected to: svaviracitā yaṃtra, but crossed out) (yaṃtra)ciṃtāmaṇivivṛti (vivṛti crossed out) samāptā.
Post-colophon: saṃ 1763 mārgaśīrṣakṛṣṇe 6 gurau śrīmanmahārājādhirāja-śrījayasiṃha devajīkasyājñayā likhitam idaṃ tulārāmeṇa.
Price on f. 1: kī –|||ı.

The **PRATODAYANTRA** composed by Munīśvara and incorporated by him in the uttarārdha of the *Siddhāntasarvabhauma* that he composed at Kāśī in 1646. For the siddhānta see *CESS* A4, 438b–439b, and A5, 314a. The *Pratodayantra* was edited with Munīśvara's own ṭīkā by S. D. Sharma, Kurali [1982]; see Y. Ohashi in *IJHS* 33, 1998, supplement pp. 188–192. The first verse is:

> gaṇeśoditaṃ yantram etat pratodam
> anāyāsakālāvabodhaṃ pravacmi |
> nijecchāvaśād iṣṭadairghyaḥ sudaṇḍo
> 'natisthūlakaḥ śiṃśivṛkṣādijātaḥ ||

Manuscript:

239. Puṇḍarīka jyotiṣa 44(a). Ff. 1–5. 10 × 21 cm. 7/8 lines. With Munīśvara's own ṭīkā. With some marginal notes.

Colophon on f. 5v: iti pratodayaṃtraṃ samāptaṃ.

The **PRATODAYANTRAṬĪKĀ** composed by Munīśvara and incorporated into his *Āśayaprakāśini*, a commentary on the *Siddhāntasārvabhauma* which he completed at Kāśī in 1650. For the *Āśayaprakāśini* see *CESS* A4, 439b–440a, and A5, 314a–314b. The *Pratodayantraṭīkā* was edited by S. D. Sharma, Kurali, [1982]. It begins: etat prasiddhaṃ pratodayaṃtraṃ gaṇeśadaivajñair asmadgurutamaiḥ kalpitam ahaṃ nirūpayāmi.

See manuscript 239.

The **YANTRARĀJARACANĀ** composed by or for Mahārājādhirāja Jayasiṃha in about 1720. See *CESS* A3, 63b–64a; A4, 97b; and A5, 117b. Edited by Kedāranātha Jyotirvid in *RPG* 5, Jayapura 1953, pp. 1–17. It begins: ādāv abhīṣṭaṃ yantraṃ dhātujaṃ dārujaṃ vā vartulaṃ kāryam.

Manuscript:

240. Puṇḍarīka jyotiṣa 10. Ff. ⟨A⟩ and 1–13. 12.5 × 28 cm. 9 lines. With some marginalia. Formerly property of Viśveśvara, the son of Rameśvara.

Colophon on f. 13v: iti śrīmanmahārājādhirājarājarājeśvaraśrībhūpālabhūpaśrījayasiṃhakāritā racanopapattivedhaprakriyā samāptā.

Ownership note by a second scribe on f. ⟨A⟩:

> rāmeśvaratanūjasya viśveśvaravipaścitaḥ ||
> yaṃtrarājavidheḥ pustaṃ jagaj jānātv idaṃ śubham || 1 ||

After this a third scribe has written:

> kāle vāridharāṇām apatitayā naiva śakyate sthātuṃ |
> utkaṃṭhitāsi tarale na hi na hi sakhi pichilaḥ paṃthāḥ || 1 ||

vyādhūya yad rasanam aṃbujalocanāyā
vakṣojayoḥ kanakakuṃbhavikāśabhājoḥ |
āliṃgasi prasabham aṃgam aśeṣam asyā
dhanyas tv ameva sacalācalagaṃdhavāha || 1 ||

utkṣiptaṃ saha kauśikasya pulakaiḥ sārddhaṃ mukhair nāsitaṃ
bhūpānāṃ janakasya saṃśayapriyā sā kaṃsam āspālitaṃ |
vaidīṃ hīmanasā samaṃ ca sahasā kṛṣṭaṃ tato bhārgava
prauḍhāhaṃ kṛtikaṃ dalena ca samaṃ tadbhagnam aiśaṃ dhanuḥ || 1 ||

Ff. ⟨Av⟩–1. Blank.

The **YANTRASĀRA** composed by Nandarāma Miśra in 1771. See *CESS* A3, 130b, and A5, 157a–157b. The first verse is:

śrīmadgokulanāthaṃ natvā vakṣye laghukriyayā |
kiṃcid yantrasamudrāt sāraṃ bālāvabodhāya ||

Manuscripts:

241. Puṇḍarīka jyotiṣa 14. Ff. 1–36, 36b, and 37–39. 11.5 × 25.5 cm. 9 lines. With some marginal notes. Copied on Sunday 1 January 1792.

F. 36 was left blank by the original scribe; a second scribe has written on it a table of the Sines (with R=60) and the reversed shadows (tangents) in digits for 1°to 90°, though entitled pratyaṃśaṃ jyākrāṃtyaṃśā vilomachāyā cāṃgulādyā. This table is imperfect; aśuddheyaṃ sāraṇī is written above it.

Ff. 36b–36bv were added by this second scribe to supply a correct table of Sines (with R=60), declinations (with ϵ=23;35,0°), and tangents measured in digits (with g=12) for 1°to 90°.

Colophon on f. 39: iti śrīmiśranaṃdarāmaviracite yaṃtrasāre upasaṃhārādhyāyaḥ samāpto yaṃ graṃthaḥ.

Post-colophon: samvat 1849 || varṣe pauṣaśukla 8 ravivāre.

F. 39v. Blank.

On f. 1 the second scribe has written an anukramaṇikā (on the left is noted the folium number, on the right the adhyāya number):

	palayaṃtraṃ 1
2	dhanuryaṃtraṃ 2
	phalakayaṃtraṃ 3
3	kapālayaṃtraṃ 4
	valayayaṃtraṃ 5
4	turīyayaṃtraṃ 6
9	sarvadeśīyaṃtraṃ 7

10 svadeśīyaṃtraṃ 8
11 anyaturīyayaṃtraṃ 9
16 laghuprakārasarvadeśīyaṃ 10
svadeśīyaṃ 11
yaṃtraṃ vinaiva dinādyānayanaṃ 12
yaṃtrarājaprakāraḥ 13
21 krāṃtivalayaṃ 14
golaracanāprakāraḥ 15
33 upaskāraracanāpra 16
golarahasyanāma 17
35 śaṃkuyaṃtraṃ 18
daṃḍayaṃtrāṇi 19
37 pratyaṃśajīvāvilomachāyājñānārthaṃ cakram 20
carārddhodayādisaṃgrahāḥ 21

242. Puṇḍarīka jyotiṣa 13. Ff. 1–14 and ⟨15⟩. 10.5 × 22 cm. 9 lines. Incomplete.
F. ⟨15⟩ ends, in line 1, in 9, 12a: carajyayā.

The **PALABHĀYANTRA** composed by Viśveśvara Mahāśabda Pauṇḍarīkayājin in about 1790/1800. (See *CESS* A5, 700b–701b.) The first verse is:

gaṇeśaṃ jānakījāniṃ natvā viśeveśvaraḥ sudhīḥ |
palabhāyantranirmāṇaṃ sugamaṃ kurute sphuṭam ||

Manuscript:

243. Puṇḍarīka jyotiṣa 32. 2 ff. 10 × 21.5 cm. 11/12 lines.
F. ⟨2v⟩ ends: tasminn eva vṛtte datvā sarveṣv apy asreṣu ghaṭyaḥ sādhanīyāḥ.

The **NALIKĀBANDHAKRAMAPADDHATI** composed by Rāmakṛṣṇa. See *CESS* A5, 453a. The first verse is:

śrīkeśavaṃ namaskṛtya rahasyapadyasaṅgrahāt |
karomi nalikābandhaṃ rāmakṛṣṇo yathākramam ||

Manuscript:

244. Khasmohor 5295. Ff. 1–8. 12 × 25.5 cm. 8 lines. Copied by Tulārāma on Tuesday 17 September 1706.

Colophon on f. 8v: iti jyotirvidrāmakṛṣṇaviracitā nalikābaṃdhakramapaddhatiḥ | samāptā.

Post-colophon: saṃ 1763 āśvinakṛṣṇa 7 bhaume likhitam idaṃ tulārāmeṇa.

Price on f. 1: kī = ||ȷ.

A text on the gnomon in 10 verses. The first is Gaṇeśa, *Grahalāghava* tripraśnādhikāra 21:

vṛtte samabhūgate tu kendrasthita-
śaṅkoḥ kramaśo viśaty apaiti |
chāyāgram ihāparā ca pūrvā
tābhyāṃ siddhatimer udak ca yāmyā ||

Manuscript:

245. Puṇḍarīka jyotiṣa 18. 1 blank folium and ff. 1 and ⟨2⟩. 10 × 21.5 cm. 9 lines. With marginal notes.

F. 1v ends in 9d: paṃkter dalaṃ syāt padaṃ | 9 |

On f. ⟨2⟩ verse 10 is found on f. 5v of manuscript 79:

mūlāvaśeṣakaṃ saikaṃ ṣaṣṭighnaṃ vikalānvitaṃ
dvisaṃguṇadviyuktena mūlenāptaṃ sphuṭībhavet || 10 ||

After this is written:

meṣādāv uttaro golaḥ tulādau dakṣiṇa(ḥ) smṛtaḥ ||

A second scribe has added lines found on f. ⟨A⟩ of manuscript 79:

chāyeṣughnākṣabhāyāḥ kṛtidaśamalavanā yamāśāḥ palāṃśāḥ ||
atha vedābdhyabdhyūnaḥ kharasahṛtaḥ śako yanāṃśāḥ ||

F. ⟨2v⟩. Blank.

A treatise on determining the ghaṭī at Jayapura from the length of a gnomon's shadow. It begins: savāījayanagare sveṣṭachāyāto ghaṭījñānam.

Manuscript:

246. Puṇḍarīka jyotiṣa 51. Ff. 1–2. 10 × 21.5 cm. 9 lines on f. 1v.

F. 2 ends, at the beginning of line 5: sveṣṭachāyāto ghaṭījñānasya patram.

F. 2v. Blank.

I. MISCELLANEOUS

The **SIDDHĀNTASAṂHITĀSĀRASAMUCCAYA** composed by Sūrya at Pārthapura in about 1543. See *CESS* A6. The first verse is not available.

Manuscript:

247. Khasmohor 5026. 1 blank folium and ff. 1–8 and 10–49 (text continuous). 12 × 33.5 cm. 5–10 lines. Ff. 14–23 also numbered 1–10; ff. 24–31 also numbered 1–8; ff. 32–40 also numbered 1–9; and ff. 41–49 also numbered 1–9. With marginalia. Incomplete (sauramāna 14–51 on ff. 1–3v; cāndramāna 1–22 on ff. 3v–4v; sāvanamāna 5–38 on ff. 4v–6; pitṛmāna 1–15 on ff. 6–7; śiśirartu 1–26 on ff. 7–8v; vasanta 1–17 on ff. 8v and 10; grīṣma 1–15 on ff. 10–11; varṣāḥ 1–9 on ff. 11–11v; śarad 1–13 on ff. 11v–12v; hema 1–9 on ff. 12v–13; bhuvana 1–63 on ff. 13–17; mahābhūjādipiṇḍotpatti 1–79 on ff. 17–22; sṛṣṭipralaya 1–91 on ff. 22–27; mlecchamatenātra bhūmisaṃsthāna 1–81 on ff. 27–30v; graharkṣamūrti 1–43 on ff. 30v–32; malamāsa 1–30 on ff. 32–34; sphuṭagativāsanā 1–9 on ff. 34–35; chedyakavicāra 1–10 on ff. 35–36; udayāstanirūpaṇa 1–19+ on ff. 36–37; grahaṇavicāra 5–21 on ff. 37–38v; ⟨candragrahaṇa⟩ 1–4 on ff. 38v–39; rāhuvicāra 1–13 on ff. 39–40; grahayutinirūpaṇa 1–4 on ff. 40–40v; and jyotiḥśāstrasyāmnāyamūlatra on ff. 41–49). Sūrya's source for the Mlecchamata was written in Persian.

F. 1 begins in 1, 14a: mṛgādāv atha karkyādau.

Colophon on f. 49: iti śrīmaddaivajñapaṃḍitasūryaviracite siddhāṃtasaṃhitāsārasamuccaye jyotiḥśāstrapurāṇavirodhaparihāro nāmādhyāyaḥ.

F. 49v. Blank.

The **BHŪGOLAPURĀṆA** in Hindī attributed to Mahādeva = Śiva. See *CESS* A5, 754a. It begins: pārabatī uvāca || oṃ svāmī bhuvamaṃḍāṇa kā kathau pramāṇa || utpati siṣṭi kā kahau baṣāṇa ketī dharatī kitā akāsa.

Manuscript:

248. Khasmohor 570. Ff. 1–11. 10.5 × 21 cm. 13 lines.

Colophon on f. 11: iti śrībhogalapurāṇe mahādevakathite sapuraṇa || samāpata ha.

The ⟨**BHŪGOLAVICĀRA**⟩ on the concept of the earth in the purāṇas and in the siddhāntas. Neither the beginning nor the end is available.

Manuscript:

249. Khasmohor 5032. Ff. 2–11. 10 × 22 cm. 11/12 lines. Incomplete.

F. 2 begins: catuḥpaṃcāśan nakṣatrāṇi caturbhujastaṃbhanibho meruḥ ekāṃtarakoṇagayor merukoṇavaśena tayor ekāṃtara evodaya iti.

F. 11v ends: atrāhus tātacaraṇāḥ yad etat yajñaliṃgasya bhagavato vāmanasya baliyāṃ cāpadeśena caraṇatalacihnair dhvajavajrāṃkuśakamalasudarśanair aṃkitam adyāpi taccaraṇāṃguṣṭa.

The **GOLADARPAṆA** or **BHĀGAVATAJYOTIḤŚĀSTRABHŪGOLAKHAGOLAVIRODHAPARIHĀRA** composed in expansion of Kevalarāma's *Bhāgavatajyautiṣayor bhūgolavirodhaparihāra* by Nandarāma in 1767. See *CESS* A3, 129a, and A5, 156b. The first verse is:

śrīmadgiridharadevaṃ
vrajajanahṛtkaṃjabhāskaraṃ vande |
ajño 'pi yatkṛpāto
vācaspatitulyatāṃ labhate ||

Manuscripts:

250. Puṇḍarīka jyotiṣa 16. Ff. 1–12. 10 × 22 cm. 9 lines. A shortened version. Copied in 1771/2.
Colophon on f. 12: iti bhūgolakhagolavirodhaparihāraḥ samāptaḥ.
Post-colophon: saṃ. 1828.
F. 12v. Blank.

251. Puṇḍarīka jyotiṣa 15. Ff. 1–18. 10.5 × 21.5 cm. 10 lines. With marginal corrections and additions by a second scribe. A longer version. Incomplete.
F. 1 begins with a passage not found in manuscript 250: s tatra karma mayasvaprakāśaśarīrāṇāṃ yakṣadevādīnāṃ vāsonakṣatrakakṣāparyaṃtaṃ nakṣatrakakṣaiva divo mūrddhā tataḥ kāṃcanī bhūmis.
Colophon on f. 18v: iti bhūgolakhagolavirodhaparihāre goladarpaṇanāmakaṃ graṃtha saṃpūrṇṇaṃ.

The **VIRODHAMARDANA** composed in 71 Sanskrit verses by Yajñeśvara Jośī at Puṇe in 1836. See *CESS* A5, 319a, and A6. The first verse is:

īśvaraṃ kevalaṃ naumi vicitrānekaśaktikam |
jagatsarjanasaṃsthānatirobhāvādikāraṇam ||

Manuscript:

252. Museum 205. Ff. 1–16. 15 × 20.5 cm. 14 lines. On English watermarked bluish paper. Each sheet is 29.5 × 20.5 cm., and is folded in half. With Yajñeśvara's own Marāṭhī ṭīkā. Copied in 1836/7.
Colophon on f. 16: iti jyotirvidyajñeśvaraviracito virodhamardananāmā graṃthaḥ samāptaḥ.

The ṭīkā ends on ff. 16–16v: yā pramāṇe yajñeśvarajośī puṇekara yāṇī prākṛtabhāṣeneṃ kele svakṛta[f. 16v]virodhamardanaprakāśa yānāve vivaraṇa samāpta.

Post-colophon: miti śake 1758.

The **VIRODHAMARDANAVIVARAṆA**, a commentary on his own *Virodhamardana* composed in Marāṭhī by Yajñeśvara Jośī at Puṇe in 1836. See *CESS* A6. It begins: hā virodhamardana yānāce navīna graṃtha va prākṛtabhāṣenetyā ce vivaraṇa puṇekara yajñeśvara jośi | yāṇī kele śake 1758.

See manuscript 252.

J. TRANSLATIONS

Theodosius wrote the *Σφαιρικά* in about the first century B.C. This was translated into Arabic as the *Kitāb al-ukarr* by Qusṭā ibn Lūqā in the late ninth century and corrected by Thābit ibn Qurra. A *Taḥrīr al-ukarr* based on this translation was completed by Naṣīr al-Dīn al-Ṭūsī in 1253.

The **UKARA**, a translation of Naṣīr al-Dīn's *Taḥrīr al-ukarr* completed by Nayanasukhopādhyāya at Jayapura in 1729. *CESS* A3 132a; A4, 122a; and A5, 159a. Edited by V. Bhaṭṭācārya as *SBG* 104, Vārāṇasī 1978. It begins in the Jaipur copy: atha ukarākhyo graṃthaḥ sāvajūsayūsakṛto likhyate || tatrā'sya trayo dhyāyā ekonaṣaṣṭitulyāni 59 kṣetrāṇi saṃti || paraṃ ca kvacit pustake ṣṭapaṃcāśa58nmitāny eva kṣetrāṇi saṃti || tatra prayojanadvayasyaikakṣetram eva kṛtaṃ || idaṃ yūnānībhāṣātaḥ arababhāṣāyāṃ abula avvāsa ahamadasyājñayā kustā viṇi lūkā bālbakvī saṃjñena tṛtīyā'dhyāyasya paṃcamakṣetraparyaṃtaṃ grathitaṃ śeṣam anyair grathitaṃ || idaṃ sābita vini kusai saṃjñena śodhitaṃ || nasīrasaṃjñena ṭīkā kṛtā || seyaṃ saṃskṛtaśabdair nibaddhyate.

Manuscript:

253. Museum 44. 1 blank folium; ff. 1–46; and 3 blank folia. 23 × 17.5 cm. 23 lines. Writing parallel to the shorter edge. Bound in cloth. Copied ⟨by Lakṣmīdhara Lekhaka⟩ on Thursday 23 October 1729. Acquired for Jayasiṃha's library in 1730.

Colophon on f. 46: ity ukaraṃ sāvajūsayūsasaṃjñasya samāptaṃ. asmin adhyāyatrayam ekonaṣaṣṭi59kṣetrāṇi saṃtīti.

Post-colophon: saṃ 1786 kā śukla 13 gurau || samāptaś cā'yaṃ graṃthaḥ.

On f. 46v is written: idam ārabībhāṣāta āvidasaṃjñaiḥ kathitaṃ na(ya)nasukhopādhyāyaiḥ saṃskṛte grathitaṃ.

Ptolemy composed the *Σύνταξις μαθηματική* in about 150. It was translated into Arabic by Isḥāq ibn Ḥunayn and corrected by Thābit ibn Qurra. On the basis of this Naṣīr al-Dīn al-Ṭūsī completed the *Taḥrīr al-Majisṭī* in 1247.

The **SAMRĀṬSIDDHĀNTA**, a translation of the *Taḥrīr al-Majisṭī* completed by Jagannātha Samrāṭ for Jayasiṃha in 1732. *CESS* A3, 57a–58a; A4, 95a; and A5, 114a. Edited by R. S. Sharma, 2 vols., Naī Dillī 1967. The first seven verses are:

gaṇādhipaṃ surārcitaṃ samastakāmadaṃ nṛṇām |
praśastabhūtibhūṣitaṃ smarāmi vighnavāraṇam ||
lakṣmīnṛsiṃhacaraṇāmburūhaṃ sureśair
vandyaṃ samastajanasevitareṇugandham |
vāgdevatāṃ nikhilamohatamopahantrīṃ
vande guruṃ gaṇitaśāstraviśāradam ||
śrīgovindasamāhvayādivibudhān vṛndāṭavīnirgatān
yas tatraiva nirākulaḥ śucimanobhāvaḥ svabhaktānayat |

mlecchān mānasamunnatān svatarasā nirjitya bhūmaṇḍale
jīyāc chrījayasiṃhadevanṛpatiḥ śrīrājarājeśvaraḥ ||
rājādhirājo jayasiṃhadevaḥ śrīmatsyadeśādhipatiś ca samrāṭ
śrīrāmapādāmbujasaktacitto yajvā sadā dānarataḥ suśīlaḥ ||
golādiyantreṣu navīnayuktipracāradakṣo gaṇitāgamajñaḥ |
satyapriyaḥ satyarataḥ kṛpālus tigmapratāpo jayati kṣamāyām ||
sa dharmapālo gaṇitapravīṇān jyotirvido golavicāradakṣān |
kārūṃs tathāhūya cakāra vedhaṃ golādiyantrair dyusadāṃ ca bhānām |
granthaṃ siddhāntasamrājaṃ samrāṭ racayati sphuṭam |
tuṣṭyai śrījayasiṃhasya jagannāthāhvayaḥ kṛtī ||
arabībhāṣayā grantho mijastīnāmakaḥ sthitaḥ |
bālakānāṃ subodhyāya gīrvāṇyā prakaṭīkṛtaḥ ||

asya granthasya trayodaśādhyāyāḥ santi | ekacatvāriṃśadadhikaśataṃ prakaraṇāni santi | ṣaṇṇavatyuttaraśataṃ kṣetrāṇi santi ||

Manuscript:

254. Pothīkhānā 183. Part I. Ff. 1–75, 75b–115, 117–144, 144b–160, 161/2, and 163–210 (text continuous). Part II. Ff. 212–285, 266 (=286) and 287–427 (text continucous). 19 × 40.5 cm. 12 lines.

The first part of the manuscript contains Jagannātha's translation of Naṣīr al-Dīn's *Taḥrīr kitāb uṣūl al-handasa* based on the Arabic version of Euclid's Στοιχεῖα. With marginal notes.

Colophon on f. 210: iti śrīrekhāgaṇite pañcadaśo dhyāyaḥ samāpto yaṃ graṃthaḥ.

The second part of the codex contains Jagannātha's *Samrāṭsiddhānta*. Spaces for diagrams left blank. Incomplete.

F. 212 begins in *Samrāṭsiddhānta* 1, 1 (vol. 1, p. 17, 7): dviguṇaḥ 7200.

F. 300. Blank.

F. 427 ends in 11, 10 (vol. 2, p. 940, 5): aṃśāṃtareṇa li.

F. 427v. Blank.

We were unable to examine manuscript 32 in the Reserved Collection, which contains the *Samrāṭsiddhānta* and the *Rekhāgaṇita*.

Al-Zarqāllu wrote several works in Arabic on the safīḥa at Toledo in about 1050 to 1075; these include the following.

The **JARAKĀLĪYANTRA** or **SARVADEŚĪYAJARAKĀLĪYANTRA** is based on a description of the universal astrolabe written in either Arabic or Persian; in its beginning it is close to the first chapter of the text published by R. Puig in *Los Tratados de Construcción y Uso de la Azafea de Azarquiel*, Madrid 1987, but afterwards diverges. It has beeen tentatively suggested that it was translated by Nayanasukhopādhyāya; see *CESS* A5, 159a–159b. Since it was incorporated by Jagannātha into the second version of his *Siddhāntakaustubha* (pp. 96–105 of the *Siddhāntasamrāṭ* edited by M. Caturveda, Sāgara

1976), the text was known in Jayasiṃha's court before 1730. The text begins: atha jarakālīyaṃtraṃ likhyate || prathamaṃ abhīṣṭaṃ vṛttaṃ kāryaṃ || tad bhāṃśāṅkitaṃ pūrvāpararekhāṅkitam ūrdhvādhararekhāṅkitaṃ ca kāryam.

Manuscripts:

255. Khasmohor 5483. Ff. 1–9, ⟨10⟩, 11–12, and ⟨13⟩, and 3 blank folia. 20 × 15 cm. 16 lines. Writing parallel to the shorter edge. Copied by a scribe whose script is similar to Kṛpārāma's.

F. 13 ends: atha svadeśalagnajñānāt anyadeśalagnajñānam āha || deśāṃtarāṃśān khalagnaviṣuvāṃśān saṃyojya lagnaṃ sādhyaṃ || yadi anyadeśaḥ svapūrvato bhavati yadi paścimataḥ tadā aṃtaraṃ kṛtvā lagnaṃ sādhyaṃ.

256. Puṇḍarīka jyotiṣa 28. Ff. 1–8. 9.5 × 32 cm. 8 lines. Ff. 4 and 8 colored yellow. With some marginalia. Incomplete.

F. 8v ends, in the middle of line 7: atha svadeśalagnajñānāt anyadeśalagnajñānam āha || deśāṃtarāṃśān svala.

Below this is written: siddhāṃtaśekhare ||

> ādye pade ⟨'⟩pacayanī palabhālpikā syāt
> chāyālpikā bhavati vṛddhimatī dvitīye ||
> akṣadyuteḥ samadhikopacitā tṛtīye
> turye punaḥ kṣayavatī tadanalpik⟨ā⟩[ṃ] ca || 1 ||
> vṛddhiṃ prayā⟨ṃ⟩tī yadi dakṣiṇāgrā
> chāyā tathāpi prathamaṃ padaṃ syāt ||
> hrāsaṃ vrajantīm atha tāṃ vilokya
> raver vijānīhi padaṃ dvi(tīyaṃ || 2 ||)

These are Śrīpati, *Siddhāntaśekhara* 4, 70–71.

Naṣīr al-Dīn al-Ṭūsī was the author of a popular treatise in Persian on the astrolabe; it is entitled *Bīst bāb dar usṭurlāb* or *Risālat al-usṭurlāb* and was probably composed in the 1240's.

The **YANTRRĀJASYA RASĀLA** called the **VĪSAVĀVA**, a translation of Naṣīr al-Dīn's *Bīst bāb*, was made by an unknown scholar; a note by a second hand at the end of Benares 81865 states: iti nayanasukhopādhyāyakṛtayaṃtrarājavicāraviṃśādhyāyī ārabītaḥ saṃskṛtanītā. If true, this would indicate that the translation was made from an Arabic version of the original. Some probability is lent to this story by the fact that the manuscript in the Jaipur Museum was copied by Kṛpārāma who copied Nayanasukha's translation of Naṣīr al-Dīn's *Tadhkira* in 1729. See *CESS* A3, 145a; A4, 125a; and A5, 159a. Edited by V. Bhaṭṭācārya as *SBG* 115, Vārāṇasī 1979. It begins: tatra paribhāṣā | tatra yantrāntargatakhaṇḍāni rekhāvṛttāni ca yāni santi teṣāṃ saṃjñā procyate.

Manuscripts:

257. Museum 42. 2 blank folia; ff. 1–28; and 2 blank folia. 22 × 16.5 cm (folia 22 × 33.5 cm folded in half). 16 lines. Bound with string to the left. Writing parallel to the shorter edge. Copied by Kṛpārāma in *ca.* 1729.

Colophon on f. 26: iti viṃśatimo dhyāyaḥ || 20 || ⊙ || yaṃtrarājasya rasāleti visavāvasaṃjñako samāpto yaṃ graṃthaḥ.

Post-colophon: li. krīpārāmeṇa.

Ff. 26v (originally) and 27–28v. Blank.

258. Puṇḍarīka jyotiṣa 12. Ff. 1–4 and 4b–23. 12 × 27 cm. 10 lines. With marginal corrections by a second scribe. Copied on Friday 28 November 1788 for the Puṇḍarīkas. Formerly property of Viśveśvara, the son of Rāmeśvara.

F. 4, line 9 ends in 3, 3 (p. 5, 11): punaḥ krāṃtivṛtta.

F. 4b, a śodhapatra, contains from there: sya tadagrimasūryāṃśam to, on f. 4bv, 3, 4 (p. 6, 2): krāṃti.

F. 4, line 10 picks up from 3, 4 (p. 6, 2): vṛttasya.

Colophon on f. 23: iti viṃśatitamā 'dhyāyaḥ | 1 || iti yavanabhāṣāyāṃ yaṃtrarājasya rasāleti vīsavāvasaṃjñakaḥ samāptaḥ.

Post-colophon: saṃvat || 18 | 45 || likhitaṃ śukravāre mārgaśīrṣasudi | 1 || idaṃ puṃḍarīkāṇām.

F. 23v. Blank.

Ownership note on f. 1:

> rāmeśvaratanūjasya viśveśvaravipaścitaḥ ||
> yaṃtrarājaparibhāṣā pustakaṃ varttate śubham || 1 ||

Naṣīr al-Dīn completed his *Tadhkira fī 'ilm al-hay'a* in Arabic in 1261; he published a revised version in 1274. This was commented on by 'Alī al-Birjandī in 1507.

The **ŚARAHATAJKIRĀ VIRJANDĪ**, a translation into Sanskrit of chapter 11 of book II of Naṣīr al-Dīn's *Tadhkira* with al-Birjandī's *Sharḥ* completed by Nayanasukhopādhyāya assisted by Muḥammad Ābidda on Tuesday 16 December 1729. See *CESS* A4, 122a. Edited by T. Kusuba and D. Pingree as *IPTS* 47, Leiden 2002. It begins: atra grahagatau yāni vastūni kaṭhinatarāni saṃdigdhāni saṃti || tadupapādakayuktayo nirūpyante.

Manuscript:

259. Museum 46. Ff. 1–56. 20.5 × 16 cm. 16/17 lines. Bound in cloth. Writing parallel to the shorter edge. Copied by Kṛpārāma. Acquired by the Jayapura library in 1730.

Colophon on f. 56v: idaṃ nasīratūsīkṛtatajkarāgraṃthasya vyākhyāne mulā alī virjaṃdīkṛte ekādaśaprakaraṇaṃ halli mā lāyana halla saṃjñaṃ samāptaṃ

|| idaṃ mahammadāviddasaṃjñai kathitaṃ nayanasukhopādhyāyaiḥ saṃskṛta-śabdair nibaddhaṃ || saṃvat 1786 śā 1651 pauṣaśukla 8 bhaume samāptim agamat.

Post-colophon: kripārāmeṇa lipikṛtaṃ.

Jamshīd al-Kāshī completed the *Zīj i Khāqānī* in Persian for Ulugh Beg at Samarqand in 1413/4.

The **VAKRAMĀRGAVICĀRA**, a translation of the chapter on planetary retrogression in the ninth chapter of al-Kāshī's *Zīj i Khāqānī*. See *CESS* A6. It begins: atha vakramārgavicāraḥ khākānījīcasya navamādhyāyastho nirūpyate || tatra nīcoccavṛttakeṃdragatyā bhūkeṃdre ye koṇā utpatsyaṃte teṣāṃ koṇānāṃ yat pramāṇaṃ saiva maṃdaspaṣṭagatir asti.

Manuscript:

260. Museum 33. Ff. 1–11 and 3 blank folia. 23.5 × 16.5 cm. 19/20 lines. Bound in red cloth with a Persian flap. Writing parallel to the shorter edge. Said to have been copied by Lakṣmīdhara Lekhaka of Sarakāra.

F. 11 ends, in the middle of line 12: evaṃ karṇebhyo gaṇitaṃ kṛtvā śīghragatayaḥ sāriṇyāṃ likhitāḥ.

F. 11v. Blank.

Ulugh Beg of Samarqand sponsored the composition of the *Zīj i Jadīd* in Persian, which was completed after 1437/8.

The **ULAKABEGĪJĪCA**, a Sanskrit translation of Ulugh Beg's *Zīj i Jadīd* or at least of its tables.

Manuscripts:

261. Museum 45. Ff. 1–77, 77b–87, and 89–100 (text continuous). 17 × 28.5 cm. Tables only: the entries are in Nāgarī numerals, the headings usually in Persian, but occasionally in Sanskrit. Brought from Surat by Nandarāma Jośi. The price was 20 1/2 rūpīs.

Statement on f. 100v: ulakabegījīca sūratate āīṃ māraphata naṃdarāma josī. kīmata bīsa rūpai ai 20S.

262. Khasmohor 5484. 10 ff. 23.5 × 15.5 cm. Long leaves folded at the top. Writing parallel to the shorter edge. The tables for the Moon only.

F. ⟨1⟩. A comparison of Ulugh Beg's lunar paramaters with those of the *Zīj i Shāhjahānī*. it begins: pākṣakeṃdras tu sarveṣāṃ mate ekaiva || ulgavegī śighrakeṃdre rāśyādi 3|20|35|0 hīnaḥ jāto śāhajahānī śīghrakeṃdraḥ ulgavegī madhyame 3|26|12|32| hīnaḥ jāto śāhajahānī madhyamacaṃdraḥ.

F. ⟨1v⟩. Table of the mean motions of the mandakendra (double elongation), śīghrakendra (anomaly), madhyama, and rāhu from A.H. 841 to 871 = A.D. 1437/8 to 1466/7.

F. ⟨2⟩. Table of mean motions for 30 to 300 years in steps of 30 and for 600, 900, and 1200 years.

Table of mean motions for the months of the Muslim calendar: muharam, saphara, ravilavala, ravila āṣira, jamādilāvala, jamādala āṣira, rajjaba, śābāna, ramajāna, śavāla. jilkāda, jilhijja, and muharram.

F. ⟨2v⟩. Table of mean motions for 1 to 31 days.

Ff. ⟨3–3v⟩. Table of mean motions for 1 to 60 hours.

F. ⟨4⟩. Deśāntara for 1° to 60° from Samakaraṃda, whose tūlāṃśāḥ are said to be 99;16.

Ff. ⟨4v–5⟩. First lunar equation for 0° to 359°.

F. ⟨5v⟩. Second lunar equation for 0° to 179° down and 180° to 359° up.

F. ⟨6⟩. Second lunar equation at closest distance.

F. ⟨6v⟩. Sexagesimal parts.

F. ⟨7⟩. Table for evection.

F. ⟨7v⟩. Sexagesimal parts reversed.

F. ⟨8⟩. Equation of daylight.

Ff. ⟨8v–9v⟩. Table of the Moon's latitude.

F. ⟨10⟩. Table for reducing the longitude of the Moon in its orbit to its ecliptic longitude.

A Sanskrit version of Ulugh Beg's star-catalogue with the longitudes increased by 4;8,41°, which is the precession for the 290 solar years between 1437 and 1726 at the rate of 1° in 70 years, which was Ulugh Beg's rate as well as that of earlier Muslim astronomers. The names of the constellations and of the individual stars are translated from Persian into Sanskrit; some Persian names and phrases also are transliterated into Nāgarī. There are nine columns:

1. Serial numbers of all the stars (1 to 1018).
2. Serial numbers of the stars in a given constellation.
3. The names or descriptions of the stars.
4. The longitudes in signs, degrees, minutes, and seconds.
5. The latitudes in degrees and minutes.
6. The direction of the latitude.

7 and 8. The magnitudes of the stars.

9. The planets associated with the stars, where appropriate.

Manuscript:

263. Puṇḍarīka jyotiṣa 20(b). Ff. 1–44 and ⟨45–46⟩. 13 × 26.5 cm.

F. 1 begins: uttarasthitalaghubhallūkamūrtitārāḥ 21 uttaraśara 21 surati dubbe asagara.

After the stars in each constellation is a summary. In these summaries are often given the opinions of vana sūphī (Ibn al-Sūfī) and of Batlamajūsa (Ptolemy).

F. 42v ends with star 951, which is the 20th in the constellation ketūrasa || manuṣyamukha || aśvajaghana || catuṣpāda (Centaurus).

Ff. 43–⟨45v⟩. The numbers in column 1 are 952 to 1018, and the numbers of the stars in each constellation are given, but the names and descriptions of individual stars generally and the summaries are omitted; columns 4, 5, 7, 8, and 9 are filled in (column 6 is left blank since all of these stars have southern latitudes). The constellations and the number of stars in each in this section are:

(Centaurus) 21 to 37

savo || vrak || huṃḍāra (Lupus) 1 to 19.

majamar || agnipātraṃ (Ara) 1 to 7.

ikalīl || chatratārā (Corona australis) 1 to 13.

kavākiv hūte janūvī || ekamatsya (Piscis austrinus) 1 to 11.

Ff. ⟨45v–46⟩. Some stars from the northern constellations—dube asagara || laghu bhalūka (Ursa minor), dubbe akabara || mahad bhallūka (Ursa maior), and tinnīn ajagara (Draco)—are repeated.

A catalogue of stars of the first, second, and third magnitudes. Their names are in Sanskrit, their longitudes are Ulugh Beg's precessed to 1726 and given in signs, degrees, and minutes; their latitudes are Ulugh Beg's with dirctions noted; and their magnitudes are in general Ulugh Beg's. The Sanskrit names of several hundred identifiable stars, at least as used in Jayapura under Jayasiṃha, are uniquely preserved in this document.

Manuscript:

264. Puṇḍarīka jyotiṣa 21. ff. 1–4 and 1 blank folium. 12 × 27 cm.

F. 1. Heading: saṃvat 1783 śāke 1648 krāṃtivṛttasthadhruvakāḥ sāyanāḥ samadhyaśarāś cālitāḥ

F. 4v. Conclusion:

puraiva pūrvair bhujanetradigbhi1022s
tulyāni bhāni praviśodhitāni ||
vedhoktayaṃtraiḥ punar evam eva
gṛhītavān vatlamayūsanāmā || 1 ||
mijastakākhye nijapustake ⟨'⟩pi
lilekha bhūyo rasamānakāni ||
tāny eva tatrādhikabiṃbam ṛkṣaṃ
yat tad vadaṃti prathamapramāṇaṃ || 2 ||
ṣaṣṭhapramāṇaṃ paramāṇubiṃbaṃ
punas tridhā syāt pratimānam etat ||
krameṇa pūrvadvitṛtīyabhedair
atha pravakṣye ⟨'⟩khilatārakotthāḥ || 3 ||

It is curious that these verses refer to Ptolemy's catalogue of 1022 stars in the *Almagest* rather than to Ulugh Beg's catalogue of 1018 stars in the *Zīj i Jadīd*, which was the compiler's source.

ʿAlī al-Qūshjī composed a *Risālah dar hay'at* in Persian for the Sulṭān Muḥammad ibn Murād at Istanbul in about 1470.

The **HAYATAGRANTHA**, a Sanskrit translation of al-Qūshjī's *Risālah dar hay'at* executed probably in the seventeenth century. See *CESS* A4, 57a–57b, and A5, 43b. Edited in a contaminated version by V. Bhaṭṭācārya as *SBG* 96, Vārāṇasī 1967. It begins: atha hayatagrantho likhyate | tatrādau saṃjñādhyāyaḥ madhye 'dhyāyadvayaṃ ante prakīrṇakaḥ.

Manuscript:

265. Museum 24. 1 blank folium, ff. 1–60, and one originally blank folium. 20 × 16 cm. 15–17 lines. Bound in cloth. Writing parallel to the shorter edge. Copied (by Ṭīkārāma according to a note on the recto of the last folium) in 1728/9.

F. 60v ends, in the margin, at the end of the text, but at a damaged edge (p. 142, 9–10): yadi chāyā ⟨pūrve tadā⟩ kivalā paści⟨me 'sti⟩ chāyā ⟨paścime⟩ tadā ⟨kivalā pūrve 'sti⟩.

Colophon on the verso of the first blank folium: iti śrīhayatasaṃjño graṃthaḥ samāptaḥ.

Post-colophon: saṃvat 1785 śāke 1650.

Farīd al-Dīn Mas'ūd ibn Ibrāhīm Dihlawī completed his *Zīj i Shāhjahānī*, whose epoch was the year of Shāh Jahān's accession to the Imperial throne, 1628, shortly before his death in Delhi on 20 October 1629. It was based on the *Zīj i Jadīd* of Ulugh Beg.

The **SIDDHĀNTASINDHU**, a translation of Farīd al-Dīn's *Zīj i Shāhjahānī* undertaken by Nityānanda at the command of Āsaf Khān and completed in about 1635. See *CESS* A3, 173b, and A5, 184a. The first verse is:

siddhaṃ sādhyam idaṃ vidher api jagad yac cetasi dhyāyate
yat smṛtvaiva purāpurāṇapuruṣo 'pīcchāvatārān dadhau |
yaj jyotiḥ paramaṃ nidhāya hṛdaye tepe tapo 'pīśvaras
tasmai maṅgalamūrtaye trivibudhārādhyaika tubhyaṃ namaḥ ||

Manuscripts:

266. Khasmohor 4960. Ff. 1–443, 3 ff., and ff. 444. Foliation in both Nāgarī and Persian. 42.5 × 27.5 cm. 28/29 lines. Writing parallel to the shorter edge. Bound as a Persian book. Some folia damaged by water. With marginal notes.

The tables end on f. 441, which was originally the last folio of the manuscript. It belonged once to the Emperor Shāh Jahān; his seal appears on f. 1.

In 1717 it belonged to Pīthīnātha; the horoscope of his son is written on f. 443; it is dated Thursday 10 October 1717 according to the inscription: saṃvat 1774 śāke 1638 tatra varṣe kārtikakṛṣṇa 1 guru ghaṭī 14 pala 18 bharaṇī ghaṭī 44|7 sīdha ghaṭī 36|7 caṃdratārīṣa 15 rātraśeṣaghaṭī 11 upari pīthīnāthagrahe putrajanma śubhaṃ. The date is comfirmed by the horoscope.

Planets	Text's longitudes (sidereal)	Computed longitudes (tropical)
Saturn	Libra	214°
Jupiter		150°
Mars	Capricorn	182(!)
Sun	Libra	208°
Venus	Scorpius	
Mercury	Libra	
Moon	Aries	176(!)
Rāhu	Virgo	183°

Acquired for 250 1/2 rūpas from Manasārāma according to a note (by the Jayapura librarian?) on f. 443: pothī jīca syājihānī vācata nityānaṃda māraphata manasārāma [ma]baidakī motatīnhī kīmata ru° 250 S patra 441. This acquisition is dated 1696 by a note on f. 444: pauthī jīca syājīhānī vā° nītānaṃda mā° manasārāma vaidakī moṭatīnhī yītī bhādavā sudī 6 dītavāra saṃvat 1753 kīmatī ru° 250 S.

Ff. 442 and 443 have mathematical rules and notes; a prastāra on f. 442. On the verso of the first unnumbered leaf after f. 443 is a geographical list in Persian in two columns of which each is divided into three sub-columns.

ṭūl wa ᶜarḍ baᶜd az buldān bi-hindūstān

al-buldān	ṭūl	ᶜarḍ
amarkūt	105;0	25;0
akbarābād	115;0	26;43
kawil	114;19	28;20
hānsī	112;25	29;45
dawlatābād	111;0	20;35
fatḥpūr sīkrī	115;+6	26;0
ajmīr	111;5	26;0
burhānipūr	118;0	20;40
ujayn	112;30	30;0
dānūl?	115;50	17;40
bījāpūr	110;20	16;32
aḥmadābād	118;40	23;15
kanbāyat	119;20	26;20

dākah bangālah	123;0	27;0
kūj bangālah	122;0	31;0
gawr bangālah	122;45	26;12
māndū	112;0	23;0
gwāliyar	114;0	26;29
aḥmadnagar	117;0	29;0?
lakhanaw	114;53	26;30
awdah	118;50	27;22
jawnpūr	119;50	26;36
siyālkūt	109;0	33;0
gawpāmaw	116;33	26;45
kurah?	118;10	26;49
mānikpūr	118;10	26;49
māhūrah	116;40	24;40
tahānīsīr	112;38	30;10
pānīpat	113;20	28;52
badā'ūn	114;59	27;32
baran	114;0	28;48
sunnām	112;22	30;30

jamlah samī wa dū? baldah sanah 16(?) shad kātib al-ᶜabd shāh maḥmūd bayādkārī qalam ...fī sanaḥ 1072. A.H. 1072 = A.D. 1661/2.

Column 2.

banāras	117;20	26;15
dabīl yuᶜanā tattah	102;30	25;10
manṣūrah qaṣbah sind	105;0	27;40
mā'aḥ az sāḥil bahr hind	102;0	19;20
nahrwālah gujarāt	102;30	22;0
sawmnāt	106;0	17;0
multān	107;35	29;40
qandahār	107;40	33;0
lahāwar	109;20	31;50
qannawj dār mulk hind	115;50	25;35
sarāndīb	130;0	10;0
tubbat	110;0	40;38
kashmīr	108;0	35;0
dihlī	113;35	28;39

A geographical list on the recto of the second unnumbered folio after f. 443:

akṣāṃśa	
laṃkāyāṃ	0|0
ādine	11

tilaṃge 18
alaṃguhyām 18|10
gaṃgāsāgara 18|20
havase 18|30
devagirau 20|34
aṃvaka saṃmā 18|5
maṇabhavasānikāyāṃ 21
makāra 1|20
śṛbhṛgukacha, naṃdapatra, vaṭapatra,
śrīstaṃbhatīrtha, dhavalake, āsīmalyāṃ 22
somanātha pattane, maṇgalāpurā 22|15
raivatakāvale 22|35
ujayinyām 23
dhārāyāṃ 23|30
siddhanagare, nomodeva, purocyaṇa, nilapurapattane 24
nalaṣṭanaṃda 5
ajamera 26|22
nāgaura 26
vāṇarasyām 26|15
laṣaṇāvatyāṃ 26|20
kaḍānagare 26|29
karṇakujve 26|35
mānikaṣṭate 26|49
tādrabhakte 27
jājnagare, udīsānagadra 27|5
ajodhyāyāṃ 27|27
gopālapura 27|29
būṃdīnagare 27|37
gopāmau 27|45
kolajalālyāṃ 28|4
kaṃpilā 28|10
śivasthāne 28|15
udyamulatāne 28|20
vatanagare 28|28
yoginīpure 28|35
daruhitaka 28|45
mena+ 39
mulatāne 29|4
....(one line lost)...
kurukṣetre 30|10
kapilasthāne, jālaṃdhare, bhunābhasabhāṇā 37|37
deyālapura 31
ramāhora bhāṃdhota 31|50
vadaṃstyāṃ 35
vūya 36|25
dāmagā 35|30

pāṃsaja	36|40
kāśmīra	37
valaka	39
samarakaṃda	40

On the verso of this same folio is a list, it appears, of recipients of copies of the *Siddhāntasindhu* (A); it is found also on f. 443v of manuscript 267, here denoted B.

A	B
jīca mūlakarī su pāliśāha kai	jīcamūlakarī su pātiśāhake
kitābaṣān nemāhi hai	kitāvaṣāṃ neṃmāhi he
1 bagāle ājama ṣa ko dai	1 vagāle ājama ṣāṃ ko daī
1 paṭanai abdallahaṣā ko dai	1 paṭaṇe avdullaha ṣāṃ ko daī
1 banarsī dai sāhava sūve ko	1 vanārasī daī sāhava mūve ko
1 dillī itakāda ṣā ko dai	1 dillī itakāda ṣāṃ ko daī
1 ujjayaṇi ṣā jadaura ko nayaśarīpha	1 ujjayaṇa ṣāṃ nadorā ko navaśerī ṣāṃ
1 burhānapura mahāvala ṣā khānākhānā	1 burahānapura mahāvata ṣā khāṃnakhāṃ
1 lahauri ujīra ṣā	1 lāhori ujīra ṣāṃ
1 kāśmīra japhara ṣā	1 kaśmīra japhara ṣāṃ
1 multāna	1 mulatāna
1 nityānaṃda	1 nityānaṃda
11 jamā	11 jamā

267. Museum 23. Ff. 1–171, 171/172, and 173–444. 45.5 × 32.5 cm. 29 lines. Writing parallel to the shorter side. Bound to the left. Ff. 341–348 bound in upside down and backwards. With marginal notes. Copied by Gaṅgārāma of Kaśmīra for Jayasiṃha on Thursday 6 April 1727.

F. 441 ends with a table entitled yogacālakāt phalaṃ.

Ff. 441v–442. Blank.

F. 442v. A prastāra.

F. 443. Mathematical texts.

F. 443v. List as on the verso of the second unnumbered folio after f. 443 of manuscript 266.

F. 444. Blank.

Date formula on f. 443:

> śubh⟨am⟩ astu sarvajagatāṃ parahitaniratā bhavaṃtu bhūtagaṇāḥ
> doṣāḥ prayāṃtu śāṃtiṃ sarvatra sukhībhavaṃtu lokāḥ || śrīḥ ||
> saṃvat 1784 vai vadi 11 guruvāre madhyāhnasamaye idaṃ
> nityānaṃdajīcapustakaṃ samāptim agamat || śudhaṃ ||
> mayedaṃ gaṃgārāmeṇa kāśmīreṇaiva vāsinā

likhitaṃ kāyakaṣṭena nityānaṃdajīcapustakaṃ 1
yādṛśaṃ pustakaṃ dṛṣṭvā tādṛśaṃ likhitaṃ mayā |
yadi śuddham aśuddhaṃ vā mama doṣo na vidyate 2

śrīrājājayasiṃhena idaṃ pustakaṃ likhāpitaṃ paropakārārthaṃ.

On f. 444v is a Hindī note indicating that Nityānanda's zīj in 444 folia was copied by Gaṅgāyaratna on *ca.* 24 May 1726: pothī hīdagī jīcakī mai pānā 444 līṣī gaṃgāyaratneṇa karkā jīca nītānaṃda vālīkī ... jeṭhasudi 5 saṃvata 1783.

268. Khasmohor 4961. Ff. 1–59, ⟨60⟩, 61–303, and ⟨304–440⟩. 45 × 33.5 cm. 24–27 lines. Writing parallel to the shorter edge. Bound to the left.

F. 28 ends (see f. 443 of manuscript 267):

śubham astu sarvajagatāṃ parahitaniratā bhavaṃtu bhūtagaṇ⟨ā⟩ḥ ||
dośāḥ prayāṃtu śāṃtiṃ sarvatra sukhībhavaṃtu lokāḥ || 20 ||

The rest of the manuscript contains tables.

269. Khasmohor 4962. Ff. ⟨2⟩, 4–48, ⟨49⟩, 50–55, ⟨56⟩, 57–63, ⟨64⟩, 65, ⟨66⟩, 67, ⟨68⟩, 69–78, ⟨79⟩, 80–87, ⟨88⟩, 89–92, ⟨93⟩, 94, ⟨95⟩, 96, ⟨97⟩, 98–279, ⟨280⟩, 281–314, ⟨315⟩, and 316–436. 37 × 25 cm. 21–30 lines. Many folia torn; heavily repaired. Writing parallel to the shorter edge. Bound to the left. Incomplete. Acquired by Jagannātha Jośī for rūpas 100.

F. ⟨2⟩ begins in 10a: svalpaprasādād.

F. ⟨2v⟩ ends: || 25 || āṇu + + + + mi taṃtraṃ yavanoktiyukti + + + rat kiṃai.

F. 4 begins: atha yavanoktamaṃgalācaraṇādilikhanaṃ || tasya viśvakarttur apāramahimā tadyogyo sti.

F. 436v ends in the star-catalogue with number 1018; there are no names, but just coordinates for stars 844 to 1018.

On the first fly-leaf is the note: pothi siddhāṃtasiṃdhukī patra 436 māraphata jośī jagannāthakī suṃ lī kimata rū 100.

The **YANTRAPRAKĀRA** composed for Mahārājādhirāja Jayasiṃha, perhaps at Delhi, in *ca.* 1729; in part it is based on Naṣīr al-Dīn's *Taḥrīr al-majisṭī*, later translated into Sanskrit by Jagannātha Samrāṭ. See *CESS* A5, 118a. Edited by S. R. Sarma, New Delhi, 1986–1987. It begins: atha śrīmahārājādhirājaviracitayantrāṇāṃ prakāraḥ likhyate. There follows an anukramaṇikā, and then: atha śrīmahārājādhirājaviracitayantrāṇi likhyante | tatrādau jayaprakāśayantraṃ likhyate.

Manuscripts:

270. Museum 31. A. 10 leaves. Ff. ⟨1v–7⟩ paginated 1–12; ff. ⟨7v–10⟩ blank. B. 22 leaves of which ff. ⟨1v–2v⟩, ⟨17–18⟩, and ⟨22v⟩ are blank. Two extra folia, C and D, are inserted after f. ⟨18v⟩. The resulting 20 leaves are paginated 14–53, of which C has pp. 42 and 43 and D pp. 44 and 45. Each pair of leaves is a single sheet folded in half. A and B: 22 × 16 cm. 25–27 lines. C: 11.5 × 23.5 cm. D: 15 × 32 cm. Writing parallel to the shorter edge. Bound to the left.

Pp. 1–12 contain from the anukramaṇikā through 1.8 (p. 21).
Pp. 14–21 contain 1.9 through 1.17 (p. 26) and part of 3.2–3.3.
Pp. 22–27 contain 2.1 through 2.9 (pp. 28–33).
Pp. 27–30 contain 3.4 (pp. 115–116).
P. 31 contains 3.5 (p. 117).
Pp. 32–36 contain the rest of 3.2 (pp. 105–110).
P. 37 contains 1.19 (pp. 27–28).
Pp. 38–40 contain the rest of 3.3 (pp. 111–114).
P. 41 contains 3.1 (p. 104).
Pp. 42–43 contain 3.6 (pp. 118–119).
Pp. 44–45 contain 3.7 (pp. 120–121).
Pp. 46–53 contain 2.10 through 2.12 (pp. 33–39).

271. Puṇḍarīka jyotiṣa 44(b). Ff. 1–21 and ⟨22⟩. 10.5 × 21.5 cm. 9–11 lines. Incomplete.

F. 14 ends. toward the end of line 2, in 1.9 (p. 22, 4): vahisthaṃ.

F. 14v. Blank.

F. 15 begins in 1.15 (p. 24, 22): paṃcadaśaghaṭikāś.

F. 17v, lines 9–10 end in 1.17 (p. 26, 17): samabhūmibhāge. In margin after this: atra trutiḥ. The text continues in 2.10 (p. 34, 13): vitribhakāle.

F. 18, lines 4–5 have the end of 2.10 (p. 34, 19–20): nakṣatranataghaṭikājñānaṃ. The text continues with 1.18 (p. 27, 2): atha sarvadeśīyakapālayaṃtrocyate.

F. 19, line 6 has the end of 1.18 (p. 27, 32): bhavatīti dik. The text continues with 2.11 (p. 34, 22): atha vitribhatulyacaṃdrasya.

F. ⟨22⟩ ends, at the beginning of line 9, in 2.11.1 Ex (p. 36, 2): svodayāṃ.

F. ⟨22v⟩. Blank.

Philippus de La Hire (1640–1718) published his *Tabulae Astronomicae* at Paris in 1702. This was reprinted at Paris in 1727. A copy of this second edition was brought to Jayapura by Father Manuel Figuereido on his return from Portugal in November 1730.

The **TABULAE ASTRONOMICAE** of Philippus de La Hire were transcribed from the Latin edition of 1727 by Joseph du Bois, who complete this task at Jayapura on 10 September 1732 according to the inscription on the title page: Conscriptae ex libro Parisiis impresso anno 1727 In Sauaipolis vel Sauai issa Pour Regia Curia Magni Reguli Sauaiyassang omnium Regulorum indiae Siri Mah Raja Seu Primario in imperio Mogolensi a me pauperculo H. R. alias Joseph du BOIS anno 1732 die 10 Septembris Gentilium vero in quorum urbe habito 6 Mensis Koar anni 1789. It begins: Usus Tabularum Astronomicarum. De Epochis seu Radicibus Vulgaribus. Dies Astronomici quibus hisce Tabulis utimur ipso currentis diei et vulgaris meridie incipiunt.

Manuscript:

272. Khasmohor 7832. Five leaves and pp. 1–29, ⟨30⟩, 31–69, 80–176, ⟨177–178⟩, 179–200, ⟨200b; crossed out⟩, 201–226, ⟨226b⟩, 227, ⟨227b⟩, and 228–275 (text continuous). Pp. 158, 188, 209, and 227b are blank. 22 × 16 cm. 24/25 lines. Writing parallel to the shorter edge. Bound in leather.

P. 138. End of the canons: Laus Deo trino et uno et Beatissimae Virgini Matri Mariae et Omnium Sanctorum Universitati per infinita saeculorum Saecula amen.

finitus fuit liber iste die 10 Septembris anni 1732 consecratum*que fuit incolae* S. Nicolai de Tolentino cui dies sacratus est in urbe *nova a Gannis con*dita Sauaissapour Domini me(?), Magni regis *Yassangh Sauai in imperio Mogoli*s omnium scientiarum Protec*tori dignis*simo *cuius vitam* Deus praelongam et incolumnem *sacro*... (italicized words are written over an early inscription in darker and larger letters).

P. 139. Tables begin. At bottom in larger hand: diferentia gentilium Caleuli substractiva 19 gradus 39 3.

P. 141 in margin in larger hand: Sauayapur Differentia a Penatibus 4′40″.

P. 142 at bottom in larger hand:

> Deli ex bononia 4. 16. 0 Subtr. Sawayaepur 26 56
> Sauaissapur 4 11 Subtr. 24 30 0
> odapur Kenauchu 27 0

P. 146. Above Tabulae VII Ascensio recta is written: jadwal (in Persian script) sāranī (in Nāgarī) Mātalee (in Roman script) maṭāli[c] (in Persian script).

P. 150 Above Ascensio recta: sva 5 sṛghaṭīḥ (in Nāgarī). Above Declinatio: krāti (in Nāgarī). Beside Canis minor Procyon: Lubdhac badhu. Beside Geminor. Caput boreale: Pritham punârvâsu. Beside Geminor. Caput australe: Punârvâsu. Beside Leonis cor: Mâgha. Beside Leonis Lucida lumbor: Purba phalguni. Beside Leonis Cauda: utra phalguni. Beside Virginis Spica: chitra.

P. 151. Beside Librae lanx australis: Bisagha. Beside Sagitarii in arcu Aust: Purbaghatt. Beside Aquarii Humerus sequens: sutibighai.

P. 153. Beside Medii Motus Solis: manayar Suray. Beside Epochae anni Christ: issa keage barsume. Beside column of centuries: issa kepisya. Above Iul.: rumi. Beside gregoriani: fringi.

P. 154. Monthly mean motion: meinalena. Daily mean motion: dinkelena. Yearly mean motion: barsunme bikre lena.

Pp. 156 and 157. Subtrahe descendendo: catena nishe. Adde ascendendo: yorna upar.

P. 161. In table of yearly motion of Moon, every 4th year has an intercalated day; above table is written Kabisa.

P. 176. Inclinatio Orbitae Lunae: śarakona (in Nāgarī).

P. 177. Laghu con.

P. 206. Stella (symbol of Mercury): bu (in Nāgarī).

P. 227. Tabula LV. Accelaratio fixarum supra Motum Medium Solis.

Pp. 228–257. Sexagesimal multiplication table: 1 to 60 multiplied by each number from 2 to 60.

At the front of the volume are 6 folia here designated by the letters A to F.

F. Av. Computation of the longitude of the Sun on 4 March 1732 at 1;22,3 hours (found to be 11^s 14;16,17°). The heading is: Locus Solis 4 Martii horae 48′ 3″ 1733.

F. B. Computation of the longitude of the Moon on 4 March 1732 at 6;48,3 hours (found to be 3^s 11;12,14°) and a repetition of the computation of the longitude of the Sun precisely as on f. Av. The heading for the Moon is: quaeritur locus ☾ die 4 Martii hora 6 min 48′ secund 3 anni 1732 bisext. A second scribe has added: Dehli. The heading for the Sun is: loc. ☉ 4 Martii 6 ho 48′ 3″ 1732 bisex.

F. Bv. de horoscopo primo cum Solis loco accipe ascensionem rectam Solis deinde abex ascensione superiore, et quatabis Summe differentia.

F. C. Title-page. The title is: TABULAE ASTRONOMICAE In quibus Solis Lunae reliquorum Planetarum Motus ex ipsis observationibus nulla adhibita Hypotesi traduntur habenturque praecipuarum Fixarum in nostro Horizonte Conspicuarum positiones. Iucundi Calculi Methodus Cum Geometrica ratione computandarum ecclipsium sola triangulorum rectilineorum Analysi exponitur. Ad Meridianum Observatorii Regii Parisiensis in quo habitae sunt observationes ab ipso Autore PHILIPO de la Hire Regio Matheseos Professore et Regio Scientiarum Academiae Socio.

F. Cv. Picture of the seal of Ioseph Dubois and an heraldic shield which was not filled in.

Sawaineopolis
Ff. D–Ev. Letter by Du Bois dated: Sawaiissapour die 10 septembris stilo novo annui Xrti D. 1732 Gentilium vero in quorum urbe habito 6 Mensis Koar anni 1789. Published by R. Mercier, *IJHS* 28, 1993, 157–166; recopied here as written:

[f. D] Lector benigne

Quo faeto nesio an fortuna impellente divinaque providentia disponente, post annos ferme 15 peregrinationum in diversis mundi Partibus, tandem in Vastissimo Indiae Orientalis Imperio quam Mogol dicitur perveni; uti postquam Majoris Consequentiae urbes peragravim in eius Metropoli cui nomen et Dehlii et Sciajanabad est appulli; et Cum Urbs ista Imperatorum Curia et totius Imperii Centrum sit, simul et peregrinātium refugium et extraneorum Concurrentium est assylus, Ibi ergo in Domu cuiusdam europei filii cui nomen Alexander Martin cuius Pater olim ex Gallia advenerat natus in Urbe Toulon in provincia

Provinciae. cuius nomen erat Iacob Martin Kan. (hoc est Princeps) iam defunctus. filii eius duo superstites, scilicet supra memoratus minor natu et Alius eo Major Ludovicus Martin nomine etiam et ipsi a Imperatore pari honore et deliciis habiti sunt et post Mortem Patris Ludovicus Imperatore imperante Patris defuncti nomen acceperat scilicet Monsieur Martin Kan minor etiam eiusdem Praecepto franghi[a] Kan. Id[e] est Europeus Princeps. hoc est clare dicendo inter Principes eos numeravit Imperator Cum stipendiis honorabilibus necnon etiam Primarii Chirurgii Oficio; igitur ergo apud dictum Alexandrum manens per aliquos menses inveni apud ipsum in eodem Loco seu prope manentem Europeum alium sabaudum Cui nomen Theodorus forestii et ipse Magnus. ingeniorans apud Imperatorem honoratissimo stypendio 200 rupiarum singulis mensibus; isti ergo duo Astronomiae non ignari quotidie in Matheseos problematibus ~~quotidie~~ tempus superfluum expendebant, ex quorum exercitio mihi cupiditas valde fuit etiam addiscere astronomiam quam antea[b] [f. Dv] Cum Astrologia eam confundentem abhorrebam. Sed Cum principium fundamentale istius Caelicae scientiae Arithmetica sit in primis in eam totis viribus discendam ingenium aplicavi. Alexander vero aut ex natura sua vel occupationum suorum Causa parum vel nihil doctrinae suppeditabat Theodorus vero libera et bona voluntate et Astronomiae et Euclidis Elementa Italico idiomate et Gallico subministravit. interea propter aliquas rationes oportuit mihi domicilium mutare et apud quemdam principem Minoris sphaerae Cui nomen Seid ferfarus Kam pro Medico eius servitium acceppi Veni Cum stipendio valde moderato necessitati id postulante ubi ferme per annum sum Moratus. interim Deus ter optimus Maximus. per vias mihi ex toto[c] ignotas duxit in servitium Cuiusdam Maximi Reguli Gentilis cui nomen SAWAY YASSANG. qui Astronomorum Princeps, seut uti Alphonsus Castellae in astriae dotem et astronomorum alimentis 400000 millia aureorum expendit ipse in eisdem stypendis singulis mensibus jam a sex annis 4000 rupiarum quae aequivalent Venetis aureis vel hollandicis vel Ungaris 1000. et in diebus augetur. ipse etiam in diversis Urbibus istius Imperii observatoria Magnifica Cum Machinis Maximis suo ingenio et si liceat dicere ferme suis Manibus fabricata, de qua re sum Ego, Deus est testis; OCularis[d] testis non semel nec bis, Machinae istae [f. E] sunt tam maximae Exempli gratia turris in trianguli anguli recti forma, cum linea aequinoctia in medio divisa usque ad minuta tertia et ad sensum divisibilis usque ad dena facta fuit ipse vero ex Cera molle suis manibus prototypum fecii⟨t⟩[e] et artificibus tradidit omnes enim ab eo sunt edocti turris[f] vero erit alta 73 pedum romanorum et modo astrolabium 12 ulnarum seu 36 Pedum romanorum modo sphaerae planae sed et dum ista scribo et modo incipitur aliud 108 pedum ei simile et multa alia &c. Iste ergo Regulus inveniens almagestum P P. Riccioli vidit et antea Cognoverat multam differentiam esse in suis Tabulis gentilium unde Persicas sciayahan olim Totius Indiae Orientalis Imperatoris jussu[g] tabulas in indorum lingua[h] Converti feci⟨t⟩[i] Cum Expensis 100000 rupiarum in quibus etiam aliqua differentia ferme ad gradum usque invenit unde quidam Pater Societatis Iesu natione Lusitanus Rector Collegii Agrae in eodem Imperio, ab eo missus fuit Europae pro quaerendo astronomo perito qui Pater ivit et redivit secumque istas quas ego descripsi tabulas Cum aliis instrumentis Mathematicis attulit donarium a

Lusitanorum Rege[j] et quidam Puer a Patre educatus India Oriundus maximo ingenio Praeditus nomine Petrus da Silva etiam Lusitania apud R. et Claris[m] Patrem Ioannē Baptistā Carbone astronomiae operam dedit, ad regulum venit Regulus valde laetus Tabulas suis Caracteribus transcribi Iussit et omnes suos astronomos Calcula per eas facere Iussit [f. Ev] Modo vero exoptat ut aliquis Parisios vel Londinum Pergat et ex fonte Astronomiam hauriat. Unde ergo. O peritissimi Caeli inspectores et Phaenomenon eius observatores vidite quo modo Usque inter Barbaros (si sic liceat vocare Pandectas istos) et nomen et hospitium doctrina vestra habeat Unde Credo Certe non dissonare veritatis si Clarissimae et Amplissimae Academiae scientiarum Parisiensi Cum Poeta possimus dicere.

Semper honos nomenque tuum laudesque Manebunt Non pigeat ergo o mi lector non pigeat inquam in fragili[k] ~~Chata~~ Charta studiis tuis Aeternum et[l] tibi Comparare nomen; et ab occidente in ultimis orientis partibus sicut D. f. lahirra sectatores et laudatores habere. Roga ergo syderum Mundique totius Conditorem Cuius armoniae est Contemplator ut ipse Caelorum Civem et te et me facere Dignetur Vale

Sawaineopolis

Sawai issapour die 10 septembris stilo novo annui Xρti D. 1732 Gentilium vero in quorum urbe habito 6 Mensis Koar anni 1789.

Tuus dum vivam
J. du Bois.

a. frankihi corr. to franghi. b. antea below line. c. valde crossed out, ex toto written above. d. testis at end of line, OCu written over this, laris at beginning of next line. e. feci. f. murris, t written over m. g. jussu above line. h. lungua. i. feci. j. Regae. k. fragi at end of line, gili at beginning of next line. l. et above line.

Ff. F–Fv. Index eorum[a] Praeceptorum in hoc libro Contentorū.

a. earum.

Pp. 258–263. Calculus Eclipsis ☾ Celebrati die 8 Iunii anni 1732 in urbe Dehli metropolis Indostani.

P. 262: Dies Civilis incipit in Medinoctio. et sic a media nocte usque ad meridiem sunt 12 horae. subtrahe eas et habebis 7 ergo eclipsis ista contigit die 8 Iunii 7 horis et 40′ 5″ post meridiem dico mediam eclipsem

---

semidiametrum umbrae terrae ——————	42–21
Latitudo lunae australis descendentis ——————	11–0
semidiametrum lunae ——————	15–35
motus horarius lunae ad solem	30–38

P. 263. Tipus Eclipsis lunae Celebrati die 8 Iunii 1732.

Diagram of the eclipse showing the shadow of the earth and the position of the Moon at 6, 7, 8, 9, and 10 P.M.

Incepit hora 5 et 49 minuta subtus terram et 6 54 minutis
dimidium seu medium eclipsis 7 horis 42 minuta
incipiet emergere ab umbra terrae 8 horis 28 minutis
finis 9 horis 33 minutis.
totalis duratio 3 horarum 44 minuta.

P. 264. An example of multiplication by the door-jamb method, first in Arabic-Persian numerals, then in European numerals.

P. 265. Another copy of the previous example of multiplication in European numerals. At the top of the page is written: Multiplicatio logistica Persarum et Mogolium.

Below the example, in crude letters, is written: kisna discipulus magistro suo se commendat die 11 octobris 1734.

Below this another scribe has written: febr. 29 1686 media nox.

P. 266. Fragment of a computation of a mean longitude; above it is written februarius. At the end is written in Nāgarī: mādadīla.

P. 267. At the top the times of the beginning and end of an eclipse:

in 3 48 26
fin 6 59 16

Below this is an incomplete horoscopic diagram, below which are two inscriptions by different hands in Nāgarī; the top one, which alone is legible, is ārāja.

Pp. 268–271. A description in Spanish of how to compute the longitude of the Sun; written by a new scribe.

P. 268 begins: quero locum solis anni 1715 dies 20 Martii hora $5 = 32'. 25''$ matutinis tempus medium Valencia.

el sobre dicho tiempo se convierte primero en astronomico tomando sus tiempos parciales comples y que da en esta forma: 1714 años febrero Completo; 18 dias 17 horas 32 min 25 seg.

para la segunda Correction se va a la Tabla 4 y se halla que la diferencia de los Meridianos de Valencia y $\overset{\text{madrid}}{\text{paris}}$ o de las tablas es 14 40 seg Con la nota s. que significa ser. substrastiva.

P. 271 ends: y se prosequira la misma operacion hasta hallar como antes la prostapheresi como en el exēplo siguiente que es el mismo que antes solo que se supone ser el tiempo dado verdadero aparente.

P. 272. A computation of the longitude of the Moon for 16 November 1651 at 9:36; written by a different scribe. At the bottom is written: M S maria de Rosario

ora.

Pp. 273–274. Blank.

P. 275 adustus felis gelidas timet aquas.

A Sanskrit prose version of de La Hire's *Tabulae Astronomicae* produced by Jayasiṃha's paṇḍitas between 1732 and 1734. Du Bois referred on 10 September 1732 on f. E of manuscript 272 to: regulus valde laetus tabulas suis caracteribus transcribi iussit; this is probably Kevalarāma's *Dṛkpakṣasāriṇī* rather than the Sanskrit prose version. See D. Pingree, *PAPS* 143, 1999, 81.

The Sanskrit prose version begins: atha ravyādigrahāṇāṃ madhyamakriyā spaṣṭakriyā ca sodākaraṇā likhyate || tatrādau abdapaśuddhyor ānayanena ahargaṇānayanam.

The *Phiraṅgicandracchedyopayogika* discussing nine geometrical figures illustrating astronomical models. This was probably drawn up after the arrival in Jayapura of Fathers Boudier and Pons in August 1734. See D. Pingree, *PAPS* 143, 1999, 81–85.

The *Phiraṅgicandracchedyopayogika* begins: tatra asarekhāṃ vyāsaṃ kṛtvā ekaṃ vṛttaṃ kāryaṃ tadupari ekaṃ sūcīkṣetraṃ kāryam.

Manuscripts:

273. Khasmohor 5292. 2 fly leaves, ff. 1–52, and 1 fly leaf. 31.5 × 22.5 cm. 23–25 lines. Writing parallel to the shorter edge. Bound in cardboard boards covered with red and white striped cloth.

Recto of first front flyleaf. A diagram of a cone instesected by a plane to create an ellipse. This is the first diagram of the *Phiraṅgicandracchedyopayogika*.

Ff. 1–39v. The Sanskrit prose version.

F. 39v. Tables XXXIII and XXXIV of the *Tabulae Astronomicae*; and the first diagram of the *Phiraṅgicandracchedyopayogika*.

Ff. 40–52v. The *Phiraṅgicandracchedyopayogika*.

F. 52v ends: iti ravigrahaṇasya gaṇite iti karttavyatā || || atrodāharaṇaṃ || ||

274. Khasmohor 5609. Ff. 33–48. 32.5 × 23 cm. 27–29 lines. Each double folium is a single sheer folded to the left. Incomplete.

F. 33 contains Table XXXVII of the *Tabulae Astronomicae*, which is found on f. 33 of manuscript 273 also.

F. 39v contains Tables XXXIII and XXXIV of the *Tabulae Astronomicae*, which are found on f. 39v of manuscript 273 also.

Ff. 40–48. The *Phiraṅgicandracchedyopayogika* without any of the diagrams.

F. 48 ends: iti ravigrahaṇasya gaṇite iti karttavyatā || || atrodāharaṇaṃ ||

Attempts to illustrate in diagramatic form de La Hire's third lunar equation, which is successfully depicted in figure 8 of the *Phiraṅgicandracchedyopayogika*.

Manuscript:

275. Khasmohor 5182. 3 sheets. A 66 × 47.5 cm; B 63.5 × 46.5 cm; and C 65 × 47 cm.

Sheets A, B, and C contain diagrams showing the earth in an orbit about the Sun; the mean and true orbits of the Moon are shown about both the earth and the Sun; and the apsidal line of the Sun is depicted. For an illustration of part of one of these see D. Pingree, *PAPS* 143, 1999, p. 83.

Sheet A also has two diagrams showing a second and a fourth situation when the equation is at its maximum.

Records of positive and negative differences in the positions of the Moon computed by means of de La Hire's tables and, presumably, observations for each of the twelve months in the Śaka years from 1784 to 1795 (AḌ. 1727/8 to 1738/9); similar records for differences between computations with the "New Tables", which may refer to the *Zīj i Jadīd* of Ulugh Beg. The entries seem to be in minutes of arc; see D. Pingree, *PAPS* 143, 1999, pp. 80–81.

Manuscript:

276. Khasmohor 5183(a). 12 ff. 55 × 43 cm. 31 lines (top for month name and year; rest for each of the 30 days in the month). Only the obverse of each folium is written upon.

F. 1 at top: laiyarasya vedhāḥ

5183(b). 12 ff. 58.5 × 38 cm. 31 lines set up as in 5183(a).

F. 1 at top: navīnasāraṇivedhāḥ.

INDICES

Concordance of library numbers and catalogue numbers

Khasmohor

157(a)	70	5079	159	5182	275	5301	42	5495(b)	177
570	248	5081	132	5183	276	5302	11	5496	115
1086	81	5082(a)	227	5187	14	5303	43	5504(a)	181
1257	221	5082(b)	195	5188	20	5305	54	5504(b)	197
1620	40	5082(d)	167	5189	19	5309	26	5504(c)	102
2954	226	5082(e)	133	5190	15	5319	75	5504(d)	103
4510	144	5082(g)	188	5195	52	5349(a)	145	5506	120
4876	216	5083	154	5197	118	5349(b)	155	5517	121
4877	219	5086	82	5198	71	5349(c)	162	5518	122
4878	220	5122(a)	83	5200	108	5349(d)	192	5519	124
4951	72	5122(b)	89	5203	56	5352	111	5520	187
4953	47	5125	128	5205	25	5353(a)	104	5530	85
4954	238	5129	210	5206(a)	237	5353(b)	183	5535	86
4955	44	5139	9	5206(b)	235	5355	88	5536	92
4958	229	5142	12	5245	32	5356	57	5570	127
4960	266	5144	61	5246	218	5358	50	5572	51
4961	268	5145	13	5247	129	5360	6	5573	16
4962	269	5174(a)	206	5249(a)	109	5362	88	5574	180
4973	231	5174(b)	160	5249(b)	125	5378	119	5576	66
4974	230	5174(c)	179	5250	175	5390	33	5580	64
4975	233	5174(d)	200	5251	178	5391	30	5583	58
5011	76	5174(f)	207	5253(a)	98	5392	28	5588	143
5013	10	5174(g)	114	5253(b)	185	5394	77	5589	46
5014	157	5174(h)	113	5254(a)	137	5403	8	5591	41
5015(a)	130	5174(i)	184	5254(b)	174	5406	63	5592	21
5015(b)	163	5174(j)	97	5254(c)	201	5408	49	5593	24
5015(c)	131	5174(k)	96	5254(d)	191	5420	156	5594	126
5016	87	5174(l)	161	5254(e)	138	5424(a)	99	5595	112
5026	247	5174(m)	160	5254(f)	198	5424(b)	101	5596	165
5030	29	5174(n)	173	5254(g)	110	5427	59	5597	105
5032	249	5174(o)	208	5254(h)	169	5429	67	5598	106
5049(a)	166	5174(p)	134	5254(i)	139	5439	31	5600	78
5049(b)	147	5174(q)	196	5254(j)	99	5446	95	5601	73
5050	158	5174(r)	190	5255	69	5469	232	5602	151
5051	141	5174(s)	160	5257	37	5479	205	5607	150
5052	100	5174(t)	135	5283	84	5480	203	5609	274
5055	68	5174(u)	182	5284	91	5481	202	7832	272
5056	107	5174(w)	186	5292	273	5483	255		
5057	55	5175(a)	136	5294	146	5484	262		
5059(a)	149	5175(b)	62	5295	244	5488	23		
5059(b)	148	5176(a)	140	5299	116	5489	22		
5060	168	5178	217	5300	211	5495(a)	176		

Museum

23	267
24	265
31	270
33	260
42	257
43	3
44	253
45	261
46	259
47	38
176	53
205	252

Pothikhānā

22	4
23	5
24	17
25	35
183	254

Puṇḍarīka jyotiṣa

1	18	34	93
2	7	35	90
3	39	36	152
4	36	37	74
5	45	38	94
6	34	39	65
7	212	40	60
9	234	41	153
10	240	42	164
11	236	43(a)	193
12	258	43(b)	223
13	242	44(a)	239
14	241	44(b)	271
15	251	44(c)	213
16	250	44(d)	214
17	80	44(e)	194
18	245	44(f)	204
19	27	44(g)	199
20(a)	222	45	224
20(b)	263	46	189
21	264	47	172
26	225	48	171
27	209	49	142
28	256	50	228
29	117	51	246
31	123	269	79
32	243	272	215
33	48		

Puṇḍarīka veda

261	2
262	1

Dates of manuscripts

10 May 1572 55
16 Nov. 1579 6
5 Mar. 1596 229
1596/7 144

1604/5 146
1606/7 49
11 Jan. 1617 230
2 Nov. 1617 231
25 June 1618 46
1625 76
1626/7 47
30 Aug. 1632 209
1 Nov. 1642 7
27 Jan. 1647 72
21 Sept. 1651 23
22 Jan. 1652 22
22 Sept. 1653 73
7 Dec. 1656 149
12 Jan. 1657 24
16 Apr. 1667 86
1674/5 104
6 Mar. 1679 38
10 Oct. 1690 36
28 Oct. 1699 50

17 May 1701 51
5 May–6 Aug. 1703 164
24 Nov. 1704 74
3 July 1705 8
5 July 1705 75
18 Aug. 1705 56
20 Nov. 1705 122
3 June 1706 25
31 Aug. 1706 21
5 Sept. 1706 76
17 Sept. 1706 244
23 Sept. 1706 232
11–26 Oct. 1706 87
5 Nov. 1706 41
9 Nov. 1706 9
14 Nov. 1706 238
16 Nov. 1706 233
30 Nov. 1706 10
8 Dec. 1706 52, 118
1 Mar. 1707 11
8 Sept. 1709 26
2 Apr. 1710 88
30 Mar. 1711 57
27 Mar. 1716 181
18 Oct. 1723 39
6 Apr. 1727 267
28 May 1727 27
1728/9 265
1729 257
23 Oct. 1729 253
15 Nov. 1730 123
4 July 1731 20
10 Sept. 1732 272
22 June 1733 77
19 Aug. 1733 28
25 Mar. 1737 66
21 Mar. 1741 78
1741/2 154
22 Nov. 1754 129
30 Jan. 1755 105
10 Feb. 1756 58
13 Feb. 1756 106
1 Sept. 1756 89
15 Feb.–20 June 1757 155
29 Sept. 1757 125
April/May 1769 99
11 May 1769 98, 175
21 Aug. 1769 59
30 Aug. 1769 176
1 Sept. 1769 178
7 Sept. 1769 67
5 June 1771 156
1771/2 250
31 Oct. 1788 60
14 Nov.–12 Dec. 1788 189
28 Nov. 1788 258
14 Dec. 1788 234
1788/9 89
1791/2 53
1 Jan. 1792 241
3 Apr. 1793 171
28 Apr. 1793 172
9 Oct. 1797 1

16 Jan. 1826 79
6 Mar. 1826 80
1836/7 252
10 July 1895 5
24 Aug. 1895 35

Authors

Arabic/Persian

European

Sanskrit

Titles

Arabic/Persian

European

Sanskrit

Scribes and Owners

Ṛddhinātha Vyāsa copied 164 on 5 May–6 August 1703.
Ṛṣaladāsa (also Ṛṣalalāla) Yati, pupil of Lakṣmīcandra Vācaka, copied 155 at Jayapura between ca. 15 February and 20 June 1757.

Ṛṣi Kaṭājī, pupil of Ṛṣi Gamvajī(?) and teacher of Ṛṣi Thaṣa(?), who copied 149 at Medanīpurī in 1656.
Kaḍva Dīkṣita, relative of Jayakṛṣṇa Dīkṣita, who owned 25.
Kamalākara Bhaṭṭa, son of Padmākara Bhaṭṭa of the Jāmadagnigotra and father of Lakṣmaṇa Bhaṭṭa, who copied 23 in 1651 and 22 in 1652.
Kāśī Dīkṣita, relative of Jayakṛṣṇa Dīkṣita, who owned 25.
Kīrttiratna Sūri, teacher of teacher of teacher of Paṇḍita Ḍūlīcandra, who copied 78 at Mulatāna in 1741.
Kṛpārāma copied 257 in ca. 1729; 259; and perhaps 255.
Kṛṣṇa copied 229 on 5 March 1596.
Keśava, father of the scribe who copied 47 in 1626/7.
Gaṅgārāma of Kāśmīra copied 267 for Jayasiṃha on 6 April 1727.
Gamvajī(?), teacher of teacher of Ṛṣi Thaṣa(?), who copied 149 at Medanīpurī in 1656.
Gulābrāya, father of Tejabhāna, who copied between 1755 and 1769.
Gokula (also Gokulanātha) Jośī, son of Śambhūnātha Puṇḍarīka Jośī, copiped 189 between 14 November and 12 December 1788; 171 on 3 April 1793; 172 on 28 April 1793; 93 for himself; and had 152 copied by Gopīnātha.
Gopīnātha, son of Rāmeśvara Vyāsa, copied 60 on 31 October 1788; 234 for Jāgeśvarajīka on 14 December 1788; 236 on 5 November 1789; and 152 for Gokulanātha.
Govardhana, son of Śrīnātha of the Nāgarajñāti, copied 146 for Dāmodara Miśra at Jīrṇadurga in 1604/5.
Govinda Kāka owned 117.
Govinda Daivajña had 27 copied by Dayārāma in 1727.

Choṭelāla copied 220.
Mahopādhyāya Jagaccandra, pupil of Vācaka Hīracandra, pupil of Bhaṭṭāraka Jayacandra Sūri of the Pārśvacandrasūrigaccha, and teacher of Lakṣmīcandra Muni, for whom Harṣāsahita Ṛṣi copied 88 at Yovanera in 1710.
Jagannātha Jośī bought 269.
Jagannātha Samrāṭ owned 36, 39, 123.
Jayakṛṣṇa Dīkṣita, son of Śrīkṛṣṇa Dīkṣita, owned 25.
Bhaṭṭāraka Jayacandra Sūri of the Pārśvacandrasūrigaccha, teacher of teacher of teacher of Lakṣmīcandra Muni, for whom Harṣāsahita Ṛṣi copied 88 at Yovanera in 1710.
Mahārājādhirāja Jayasiṃha had Tulārāma copy 41, 52, 76, 87, 118, 232, 233, 238, and 244 in 1706; 11 in 1707; and 26 in 1709; had Nātha copy 9, 10, and 21 in 1706; and had Gaṅgārāma copy 267 in 1727.
Jāgeśvarajīka had Gopīnātha copy 234 in 1788.
Jīva Dīkṣita, relative of Jayakṛṣṇa Dīkṣita, who owned 25.
Jīvarāja Dīkṣita, relative of Jayakṛṣṇa Dīkṣita, who owned 25.
Paṇḍita Joga, father of Harinārāyaṇa Praśnorā, who copied 28 and 77 in 1733.
Joseph du Bois copied 272 in 1732.

Ṭīkārāma copied 265 in 1728/9.
Paṇḍita Ḍūlīcandra, pupil of Paṇḍita Sadāsukha Gaṇi, pupil of Darśanasundara Gaṇi, pupil of Kīrttiratna Sūri, copied 78 at Mulatāna on 21 March 1741.

Tulārāma copied for Jayasiṃha: 76 on 5 September 1706; 244 on 17 September 1706; 232 on 23 September 1706; 87 on 11 and 26 October 1706; 41 on 5 November 1706; 238 on 14 November 1706; 233 on 16 November 1706; 52 at Amadāvāda on 8 December 1706; 118 on 8 December 1706; 11 on 1 March 1707; and 26 on 8 September 1709.

Tejabhāna (also Tejabhānu and Tejobhānu), son of Gulābrāya, copied 105 on 30 January 1755; 58 for himself on ca. 10 February 1756; 106 on 13 February 1756; 125 on 29 September 1757; 99 in April/May 1769; 175 on 11 May 1769; and 98 probably at Mulatāna on 16 May 1769.

Ṛṣi Thaṣa(?), pupil of Ṛṣi Kaṭājī, pupil of Ṛṣi Gamvajī(?), copied 149 at Medanīpurī on 7 December 1656.

Dayārāma copied 27 for Govinda Daivajña on 28 May 1727.

Darśanasundara Gaṇi, pupil of Kīrttiratna Gaṇi and teacher of teacher of Paṇḍita Ḍūlīcandra, who copied 78 at Mulatāna on 21 March 1741.

Dāmodara Miśra had Govardhana copy 146 at Jīrṇadurga in 1604/5.

Devarṣi Jyotiṣī copied 144 for Lakṣmīcandra at Sāgānayari in 1506/7.

Dhanarāja, son of Vidyāvinoda Mathena, copied 51 for himself on 17 May 1701.

Dhaneśvara Pauṇḍarīka had Śyāmasundara copy 80 on 6 March 1826; and owned 225 and 228.

Nātha Bhaṭṭa, son of Viṭṭhala Bhaṭṭa of the Medapāṭhajñāti, a resident of Rūpapura in Gujarāta, copied for Jayasiṃha: 21 on 31 August 1706; 9 on 9 November 1706; and 10 on 30 November 1706.

Padmākara Bhaṭṭa of the Jāmadagnigotra, grandfather of Lakṣmaṇa Bhaṭṭa, who copied 23 in 1651 and 22 in 1652.

Pāṇḍai(?) copied 55 on 10 May 1572.

Pīthīnātha owned 266 in 1717.

Puruṣottama Vyāsa of the Śrīmālījñāti, grandfather of Śrīdatta, who copied 230 at Jodhapura on 11 January 1617; 231 at Śuddhadantī on 2 November 1617; and 46 for himself at Jodhapura on 25 June 1618.

Premarṣi copied 104 at Medanarapura in 1674/5.

Mahopādhyāya Mativilāsa of the Maladhāragaccha, teacher of teacher of Sahajasundara, who copied 6 at Argalapura in 1579.

Manasārāma owned 266 in 1696.

Mitrarāma Bhaṭa copied 117.

Rāmadatta Rāma copied 53 in 1791/2.

Rāmabhaṭa copied 1, probably at Kāśī, on 9 October 1797.

Rāmeśvara Puṇḍarīka, father of Viśveśvara, who owned 36, 39, 48, 60, 65, 90, 123, 212, 236, 240, 258.

Rāmeśvara Vyāsa, father of Gopīnātha, who copied 60 in 1788; 234 for Jāgeśvarajīka in 1788; 236 on 1789; and 152 for Gokulanātha.

Lakṣmaṇa Josyā owned 22.

Lakṣmaṇa Bhaṭṭa, son of Kamalākara Bhaṭṭa, son of Padmākara Bhaṭṭa of the Jāmadagnigotra, copied 23 on 21 September 1651 and 22 on 22 January 1652.

Lakṣmīcandra had Devarṣi Jyotiṣī copy 144 at Sāgānayari in 1506/7.

Lakṣmīcandra Muni, pupil of Mahopādhyāya Jagaccandra, pupil of Vācaka Hīracandra, pupil of Bhaṭṭāraka Jayacandra Sūri of the Pārśvacandrasūrigaccha, had Harṣāsahita Ṛṣi copy 88 at Yovanera in 1710.

Lakṣmīcandra Vācaka, teacher of Ṛṣaladāsa Yati, who copied 155 at Jayapura in 1757.

⟨Lakṣmīdhara Lekhaka⟩ copied 253 on 23 October 1729; and probably copied 260.

Līchamīnārāyaṇa copied 5 on 10 July 1895; 35 on 24 August 1895; 4; and 17.
Līlādhara copied 56 at Avantikā on 18 August 1705.
Vallabha Bhaṭṭa copied 8 at Avantīpurī on 3 July 1705.
Vāsudeva copied 72 for his son at Kāśī on 27 January 1647.
Viṭṭhala Dīkṣita copied the original of 76 in 1625.
Viṭṭhala Bhaṭṭa of the Medapāṭhajñāti, a resident of Rūpapura in Gujarāta, father of Nātha Bhaṭṭa who copied for Jayasiṃha; 9, 10, and 21 in 1706.
Vidyāvinoda Mathena, father of Dhanarāja, who copied 51 in 1701.
Vaijanātha Jośī owned 74.

Śaṅkara (also Śaṅkaradāsa) Vyāsa, son of Puruṣottama Vyāsa of the Śrīmālījñāti, father of Śrīdatta who copied 230 at Jodhapura in 1617; 231 at Śuddhadantī in 1617; and 46 at Jodhapura in 1618.
Śambhunātha Puṇḍarīka Jośī, father of Gokulanātha Jośī, who copied 189 in 1788; 171 and 172 in 1793; and 93; and had 152 copied by Gopīnātha.
Śiva Daivajña owned 84.
Śyāmasundara copied 80 for Dhaneśvara Pauṇḍarīka on 6 March 1826.
Śrīkṛṣṇa Dīkṣita, father of Jayakṛṣṇa Dīkṣita, who owned 25.
Śrīdatta, son of Śaṅkara Vyāsa, son of Puruṣottama Vyāsa of the Śrīmālījñāti, a resident of Śubhaṭapura, copied 230 at Jodhapura on 11 January 1617 during reign of Sūryasiṃha; 231 at Śuddhadantī on 2 November 1617; and 46 at Jodhapura on 25 June 1618.
Śrīdatta Jyotirvit, father of Satīdāsa, who owned 24 in 1657.
Śrīnātha of the Nāgarajñāti, father of Govardhana, who copied 146 for Dāmodhara Miśra at Jīrṇadurga in 1604/5.
Śrīnātha copied 209 on 30 August 1632.
Śrīvatsa Dīkṣita, relative of Jayakṛṣṇa Dīkṣita, who owned 25.
Satīdāsa, son of Śrīdatta Jyotirvit, owned 24 on 12 January 1657.
Paṇḍita Sadāsukha Gaṇi, pupil of Darśanasundara Gaṇi, pupil of Kīrttiratna Gaṇi, and teacher of Paṇḍita Ḍūlīcandra, who copied 78 at Mulatāna on 21 March 1741.
Santoṣarāma copied 75 on 5 July 1705.
Saravaṇa Gaṇi, pupil of Mahopādhyāya Mativilāsa of the Maladhāragaccha and teacher of Sahajasundara, who copied 6 at Argalapura in 1579.
Sahajasundara, pupil of Saravaṇa Gaṇi, pupil of Mahopādhyāya Mativilāsa of the Maladhāragaccha, copied 6 for himself at Argalapura on 16 November 1579.
Sūrjabhāṇa copied 176 on 30 August 1769 and 178 on 1 September 1769.
Sūryabhānu owned 67.
Senāpati owned 73.
Harikṛṣṇa copied 73 on 22 September 1653.
Harinārāyaṇa Praśnorā, son of Paṇḍita Joga, copied 77 on 22 June 1733 and 28, probably at Kāśī, on 19 August 1733.
Harṣāsahita Ṛṣi copied 88 for Lakṣmīcandra Muni, pupil of Mahopādhyāya Jagaccandra, pupil of Vācaka Hīracandra, pupil of Bhaṭṭāraka Jayacandra Sūri of the Pārśvacandrasūrigaccha, at Yovanera on 2 April 1710.
Vācaka Hīracandra, pupil of Bhaṭṭāraka Jayacandra Sūri of the Pārśvacandrasūrigaccha, teacher of teacher of Lakṣmīcandra Muni, who had Harṣāsahita Ṛṣi copy 88 at Yovanera in 1710.

Other Persons

Families, Gacchas, Gotras, Jātis, and Jñātis

Āñcalikagaccha 66
Bhaṭṭa 8, 9, 10, 22, 23
Bṛhadgaccha 229, 231, 232, 234
Dīkṣita 25, 27, 76
Jāmadagnigotra 22
Jośī 74, 152, 189, 252, 261, 269
Josyā 22
Maladhāragaccha 6
Mathena 51
Medapāṭhajñāti 9, 10, 21
Mleccha 247
Nāgarajñāti 146
Pārśvacandrasūrigaccha 88
Pauṇḍarīka, Puṇḍarīka 36, 39, 48, 60, 65, 80, 90, 93, 123, 212, 225, 228, 236, 240, 243, 258
Praśnorā 28, 77
Śrīmālījñāti 230, 231
Tughlug 229–235
Vyāsa 46, 60, 164, 230, 231, 234, 236

Toponyms

European (from 272)

Agra
Bononia
Dehli, Deli
Gallia
Gallicus
Hollandicus
India
Indostanus
Indus
Italicus
Kenauchu (Kannauj?)
Londinum
Lusitania
Lusitanus
Madrid
Mogol, Mogoles
Odapur (Udaipur)
Paris
Parisiensis
Parisii
Persae
Persicus
Portugal
Provincia
Romanus
Sauai issa Pour, Sauaissapur (Savāī Īśapura)
Sauaipolis
Sauayapur
Sawai neopolis
Sawayaepur
Sciajanabad (Shāhjahānābād)
Toulon
Ungarus
Valencia
Venetus

Sanskrit

General Index